Selected Organic Syntheses

A Guidebook for Organic Chemists

Ian Fleming

*Assistant Director of Research in the Department
of Organic and Inorganic Chemistry, Cambridge*

Fellow of Pembroke College, Cambridge

A Wiley–Interscience Publication

JOHN WILEY & SONS

London · New York · Sydney · Toronto

Library of Congress catalog card number 72-615

ISBN 0 471 26390 7 Cloth bound

ISBN 0 471 26391 5 Paper bound

*Typeset in Northern Ireland at The Universities Press, Belfast
and printed in Great Britain at The Pitman Press, Bath*

'Of course, men make much use of excuses for activities which lead to discovery, and the lure of unknown structures has in the past yielded a huge dividend of unsought fact, which has been of major importance in building organic chemistry as a science. Should a surrogate now be needed, we do not hesitate to advocate the case for synthesis'.

R. B. Woodward in *Tetrahedron*, **19**, 248 (1963).

'The synthesis of substances occurring in Nature, perhaps in greater measure than activities in any other area of organic chemistry, provides a measure of the condition and power of the science. For synthetic objectives are seldom if ever taken by chance, nor will the most painstaking, or inspired, purely observational activities suffice. Synthesis must always be carried out by plan, and the synthetic frontier can be defined only in terms of the degree to which realistic planning is possible, utilizing all of the intellectual and physical tools available. It can scarcely be gainsaid that the successful outcome of a synthesis of more than thirty stages provides a test of unparalleled rigor of the predictive capacity of the science, and of the degree of its understanding of its portion of the environment. Since organic chemistry has produced syntheses of this magnitude, we can, by this yardstick, pronounce its condition good'.

R. B. Woodward in A. R. Todd (Ed.), *Perspectives in Organic Chemistry*, Interscience, New York, 1956, p. 155.

'The synthetic chemist is more than a logician and strategist; he is an explorer strongly influenced to speculate, to imagine, and even to create. These added elements provide the touch of artistry which can hardly be included in a cataloguing of the basic principles of Synthesis, but they are very real and extremely important. Further, it must be emphasized that intellectual processes such as the recognition and use of synthons require considerable ability and knowledge; here, too, genius and originality find ample opportunity for expression.

The proposition can be advanced that many of the most distinguished synthetic studies have entailed a balance between two different research philosophies, one embodying the ideal of a deductive analysis based on known methodology and current theory, and the other emphasizing innovation and even speculation. The appeal of a problem in synthesis and its attractiveness can be expected to reach a level out of all proportion to practical considerations whenever it presents a clear challenge to the creativity, originality and imagination of the expert in synthesis'.

E. J. Corey in *Pure Appl. Chem.*, **14**, 30 (1967).

Contents

The names given below are those of the senior authors responsible for the syntheses. It should be emphasised that they are not responsible for this presentation of their work, and that any errors of interpretation and emphasis are those of Ian Fleming.

Synthesis

My main aim in bringing together a number of important organic syntheses in this book has been to illustrate the reactions which every organic chemist must know.

Organic chemistry is now taught within a mechanistic framework based on the molecular orbital theory of bonding. Chemists thirty or forty years ago were faced with a large body of fact, which was organised largely along functional group lines and required a good deal of memorising. There was very little understanding of mechanism, but because chemists have kept trying to fit these facts into a scheme, the theory of mechanisms has developed into a conceptual framework which defines both the way we think about chemistry and, naturally, the way it is taught. The very large number of known reactions are now seen to belong to a relatively smaller number of reaction types, and principle is taught in place of bare fact.

Yet, even now, arguments adduced from the theory of mechanism are not powerful enough on their own to predict what could, or ought, to happen when a particular pair of reagents or functional groups are brought into combination; if a chemist is to be properly equipped, principle must be supported by a fairly extensive knowledge of the reactions of organic compounds. For example, it is not enough to know that enolate ions are nucleophilic, without knowing just how powerful an electrophile needs to be for reaction to occur between it and an enolate ion. It is still necessary to know a lot of facts; and it is some of these facts which are illustrated in this book.

To some extent, I have placed the very large number of reactions mentioned in their mechanistic framework, stressing the carbon–carbon bond forming reactions; but the main aim has been to illustrate the occasions on which the reactions learned by every organic chemist are put into practice: to show, for example, how Claisen condensations are used to make more complicated molecules than just acetoacetic ester itself.

The examples of synthesis in this book will not teach you, except by example, how to do a synthesis, but they will bring to your attention many of the factors which must be borne in mind in planning a synthesis, and will illustrate the solutions to some typical problems.

The syntheses are arranged in a roughly chronological order. I do not believe that a practising chemist needs to have a great sense of the history of a developing subject; interest in such things is properly academic. I have chosen a chronological approach simply because it is an arrangement which lets the reader start with relatively simple syntheses and encounter works of increasing complexity as he progresses; the reactions will go from those taught in all elementary courses to ones specifically designed for the synthesis in question.

The outline of each synthesis is separately presented in structural formulae,* in a way familiar to all organic chemists. I discuss in the text† those matters which are general to synthesis at the point where they first occur, but I have chiefly been concerned to amplify the matter presented as structural formulae by drawing attention to the kinds of reaction taking place, their selectivity and their stereo-chemistry.

In making my choice of syntheses I have been influenced by many factors besides those affecting my main aim of including as large a range of reactions as possible. I particularly hope, in the course of this book, to illustrate the general principles attaching to synthesis; to illustrate as large a range as possible of compounds from the natural product field; and, perhaps most importantly, to demonstrate the intellectual achievement of the masters of the art.

A few general essays have been written about the activity of synthesis in organic chemistry. In reviewing these now we shall cover, at least in outline, some of the more general aspects of synthesis. A forceful and challenging case[1] has been made by R. B. Woodward for the value of synthesis as an academic pursuit in addition to its more obvious purposes in aiding structure determination and providing material for medicinal use. The strategy of organic synthesis has been discussed[2] by E. J. Corey in accounts of how syntheses may be planned—essentially by taking molecules apart mentally, using the reverse of the reactions which might be employed to put them together. Finally, Corey has extended[3] this activity and refined its logic to the point where a computer can assist in the 'dislocation' of the target molecule by rejecting many unworkable routes and leaving a number, although a large number, of possible ones for human evaluation. Such computer-aided assessment of the task involved in a particular organic synthesis is both more complete than unaided thinking is likely to be and is also a first step in what is obviously the right direction. The achievement of this step has been made possible by the invention of the necessary hardware and software needed if structural dia-grams (the universal language of organic chemistry) are to be presented and absorbed as a connection table by the computer. All the computer does is to be more thorough than most organic chemists will be. The actual thinking is, of course, the same as that involved in the planning of any synthesis, this planning is, in its turn, much assisted by that recognition of reaction types which mechanistic insight has provided.

---

* Using structural formulae labelled with bold-face arabic numerals on a grey background.
† Using structural formulae labelled with roman numerals.

Let us examine a very simple molecule (I), and see how we might set about synthesising it.

(I)

In planning a synthesis of a molecule such as I, we look first at the array of functional groups and examine this array for any features which we may recognise as being derived from a known reaction type. In this case it is particularly obvious that the β-hydroxyketone system could be set up by an aldol reaction. Thus our first step is to look at the molecule (II) which would cyclise to I if the desired aldol condensation were to take place. The notional step of reversing the aldol reaction is known as *dislocation*, and is conveniently represented on diagrams by a double arrow $\Rightarrow$.

(I)   (II)

We now have a new target molecule (II) to examine. A further simplification of the synthetic task can often be made by recalling the interconversions which functional groups can be expected to undergo. Thus the 5-ketone group of II could be derived from a nitro compound (III) by several routes. With this

(II)   (III)

and several other possibilities in mind, we can now see that another sensible dislocation can be made:

(III)   (IV)   (V)

This dislocation takes advantage of our knowledge that in the nitroalkane (V) we have a potentially nucleophilic carbon atom at C-5, and that the unsaturated ketone (IV) is electrophilic at C-4. The Michael reaction of these two components will be catalysed by base, which is needed to remove the proton next to the nitro group:

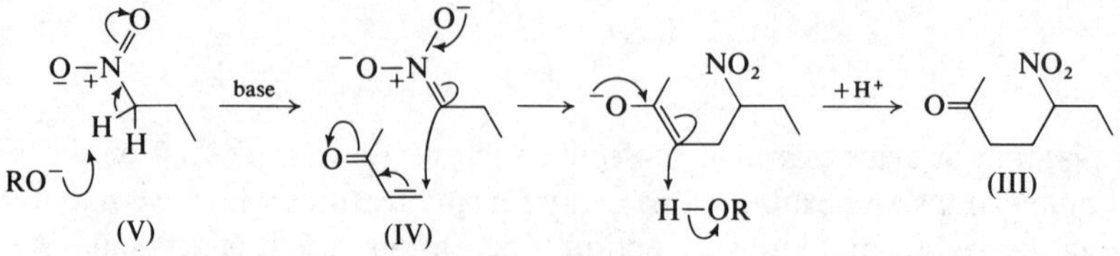

Thus, by 'transforming' the keto group of II to the nitro group of III, we have changed an electrophilic carbon atom (needed for the aldol reaction) to a nucleophilic carbon atom (needed for the Michael reaction). We have now generated a potential synthetic route, because the compounds IV and V are readily available.

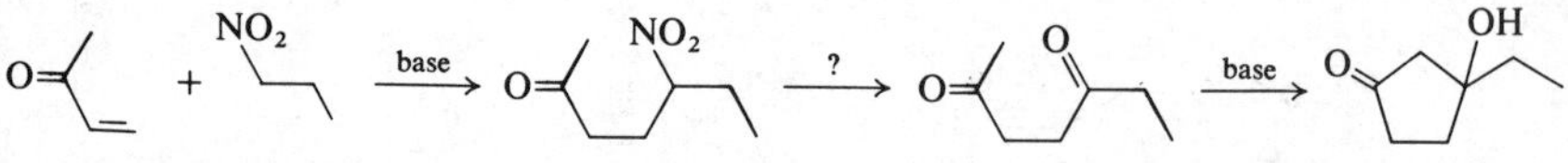

We must next examine each of the steps and the intermediates for flaws—that is, for features which might lead the compounds to react in different ways from the way we want them to.

The first thing we would do is to find out how much of this sequence is already known. We would find that the step IV + V has not actually been done, but that it has been done[4] with both the lower and higher homologues of V. We would therefore feel that that step at least was going to be easy. We would find also what reaction conditions were used with the homologues of V. They will be the obvious ones to try in the first instance, if we do eventually decide to put this route into practice. None of the sequences for converting III to II has been tried in this case, so we would have to draw from slightly more remote analogies. We would find, for example, that the acid (VII) had been prepared[5] from the nitro compound (VI) by a Nef reaction, and we might hope that the keto group in III

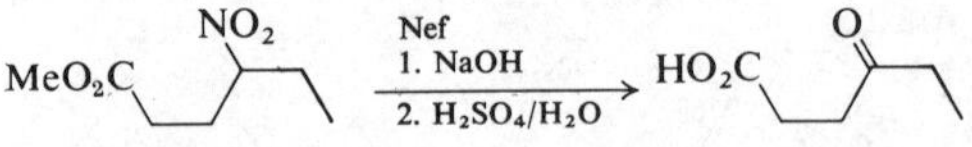

would survive these conditions. If it proved to be troublesome, we might then essay one of the other routes. No doubt one of these would be successful[6] and we would then have only one further step to go. But it is this step, the conversion of the diketone (II) to the β-hydroxyketone (I), which we can expect to be the most troublesome. For there are two likely β-hydroxyketones (I and VIII) which can be produced from the diketone (II) when it is treated with base. Some other possible products, resulting from enolisation in the other direction at each carbonyl group, are less likely, because they involve the formation of three-membered

rings; and it is known that β-hydroxycyclopropylketones such as IX and X open under the influence of base—that is, the aldol reaction goes in reverse in these cases. So we can expect two products from the aldol reaction on II and can expect that we will be obliged to separate out the one we want from the other one. Nevertheless the simplicity and directness of the route we have been examining might well make us regard it as a promising one. We can always hope—and analogy in the literature would support that hope—that we would be able, by judicious choice of reagent, to influence the ratio of the two products in the final step. The fact that aldol reactions can be catalysed by acid, by mild base and by secondary amines, and the fact that there is a considerable range of bases with differing steric requirements to choose from, would make us particularly optimistic.

We have so far examined only one route to the cyclic ketone (I). There are several others which would have to be examined and assessed in much the same way. One possibility, for example, is to make the protected derivative (XII) of the cyclic diketone (XI) and then treat it with the ethyl Grignard reagent. This

route, on further examination, would be found to suffer from the disadvantage that the cyclic diketone (XI) is not as easy to make as its simple structure might lead one to expect.

When we have examined all the routes we can think of, or all the routes suggested by Corey's computer, we would have to decide, finally, which route to embark upon. It is at this point that experience is combined with intellect to produce the final plan. Usually one other factor, which is not apparent in such a simple example as the one above, has to be incorporated in the overall design. That factor is flexibility: it is a rare synthesis which goes entirely according to plan.

In the simple example above, we have seen some part of the reasoning which has to go into the preliminary work of designing a synthesis. Careful analysis of all the factors which make each of the possible routes more or less favourable is of inestimable value. It enables one, not only to choose the most promising route, but also to anticipate by-products. Such careful preliminary work on paper and in the library pays for itself many times over; for there is nothing more disappointing and futile than rediscovering what is already known and well documented. From now on the syntheses described in this book will show only the routes which were brought to successful conclusion. They are, of course, all much more of a challenge than the simple example discussed above. The problems created by the 'diabolic concatenation of reactive groupings', the 'plethora of asymmetric centres', the 'repetitious monstrosities', and the 'very fugitive'[1] have been solved in ways which excite our admiration and reward our attention.

References

1. R. B. Woodward, in A. R. Todd (Ed.), *Perspectives in Organic Chemistry*, Interscience, New York, 1965, p. 155–184.
2. E. J. Corey, *Pure Appl. Chem.*, **14**, 19 (1967); E. J. Corey, M. Ohno, R. B. Mitra and P. A. Vatakencherry, *J. Amer. Chem. Soc.*, **86**, 478 (1964).
3. E. J. Corey and W. T. Wipke, *Science*, **166**, 178 (1969), E. J. Corey, *Quart. Rev.*, **25**, 455 (1971).
4. H. Schechter, D. E. Ley and L. Zeldin, *J. Amer. Chem. Soc.*, **74**, 3664 (1952); H, Feuer and R. Harmetz, *J. Org. Chem.*, **26**, 1061 (1961).
5. M. C. Kloetzel, *J. Amer. Chem. Soc.*, **70**, 3571 (1948).
6. After these words were set up in type, this very sequence (IV + V → III → II) was reported by J. E. McMurry and J. Melton, *J. Amer. Chem. Soc.*, **93**, 5309 (1971).

Synthesis in the Nineteenth Century

It is impossible to find a clear beginning to the story of organic synthesis. The very early production of alcohol and sugar in relatively pure form clearly represents only the purification of natural products. The preparation of ether, in the sixteenth century, and ethyl nitrate and chloride shortly after, represent the first derived products. Wöhler's famous preparation of urea[1] from ammonia and cyanic acid in 1828 was the first bridge between the organic and inorganic worlds; but it was not so clear-cut a distinction at that time as one might think.[2] The vitalist theory—that organic compounds, present in animals and plants, were somehow different and incapable of being prepared from matter of mineral origin—was not universally held; it crumbled in gradual steps and was convincingly crushed only by the synthesis of acetic acid[3] by Kolbe in 1845. Perhaps this achievement marks the beginning of our story. It is a multi-step synthesis, starting from the indisputably inorganic—the elements—and culminating in what was universally regarded as organic—acetic acid.

$$C \xrightarrow{\text{FeS}_2} CS_2 \xrightarrow{\text{Cl}_2} CCl_4 \xrightarrow[\text{tube}]{\text{red hot}} Cl_2C{=}CCl_2$$

$$\downarrow \text{sunlight, water}$$

$$CH_3CO_2H \xleftarrow{\text{electrolysis}} CCl_3CO_2H$$

acetic acid

It is interesting, too, in that successive steps in the total synthesis were not developed chronologically: the preparation of carbon disulphide from the elements had been discovered in 1796 (Lampadius) and the conversion of trichloroacetic acid to acetic acid in 1842 (Melsens) (though Kolbe actually used an electrolytic reduction). The contribution of Kolbe's work was that, as so often happens in synthesis,.it filled in the gap, though it is most unusual in doing so entirely with compounds which were already known. The last reaction to be discovered (which therefore formally effected the synthesis) was the conversion of carbon tetrachloride to tetrachloroethylene.

Up to this time the range of reactions available was small and, even more inhibiting, the concept of the structure of organic compounds was vague and various. The introduction of the correct view of structure by Kekulé and by

Couper in 1858, and the clarifying of much of the existing organic chemistry by Butlerow, by Cannizzaro and by others shortly afterwards made it possible for chemists to discover in the second half of the century many of the most important and common organic reactions. The value of interconverting compounds by synthetic sequences was most explicitly championed by Berthelot, who contributed greatly to the demonstration of these interrelationships. It was he who established the word synthesis and defined its unifying function.

'Thus synthesis extends its conquests from the elements up to the domain of the most complicated substances without our being able to assign any limit to its progress. Indeed, if we envisage in our minds the almost infinite number of organic compounds, from the substances which our art knows how to produce, such as [hydrocarbons], alcohols, and their derivatives, up to those which still exist only in nature, such as the sugars and the nitrogenous principles of animal origin, we pass from one term to the other by insensible degrees, and we cannot see any absolute barrier or break which we may with any appearance of certitude fear to find unsurpassable.'[4]

We may note as landmarks: (1) The synthesis of t-butanol[5] by Butlerow in 1863— one of the first compounds to be talked about before it had been made;

$$Me_2Zn \xrightarrow{\text{COCl}_2 \text{ or MeCOCl}} Me_3COH$$

t-butanol

(2) The (fortuitous) production by Perkin,[6] of the first commercially useful synthetic dye, mauve (a mixture of **1** and **2**), in the course of work which started as a misguided attempt to synthesise quinine by the oxidation of allylaniline;

present as impurities in
the commercial aniline
used by Perkin

(1) pseudomauveine   "mauve"   (2) mauveine

(3) The synthesis by Graebe and Liebermann[7] of alizarin (**3**), a natural dyestuff, on a commercial scale from 1869.

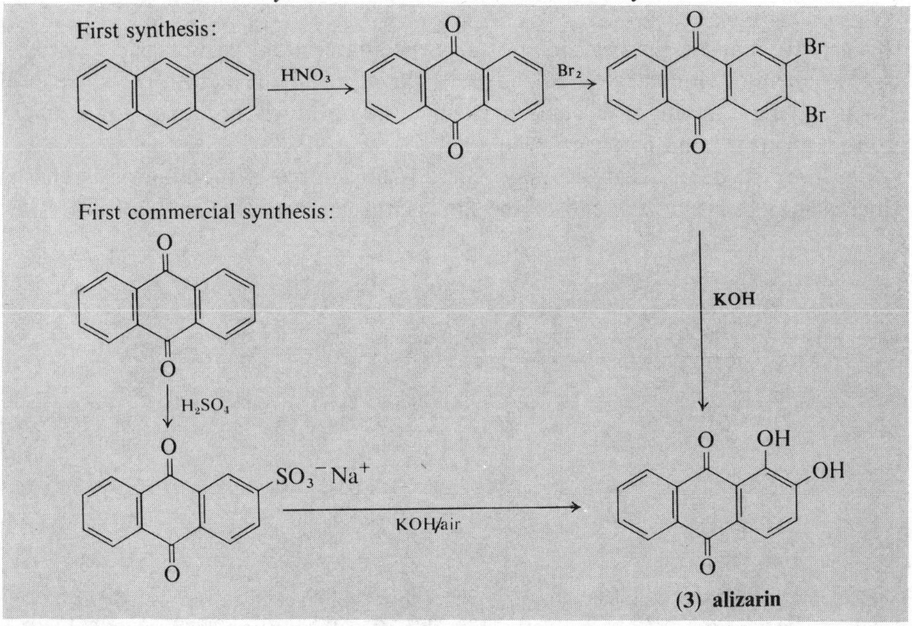

First synthesis:

First commercial synthesis:

(3) alizarin

The success of this last venture spurred onwards the work on indigo (4) which, was first synthesised in 1878 by Baeyer,[8] with commercial production, by a different route,[9] beginning in 1897.

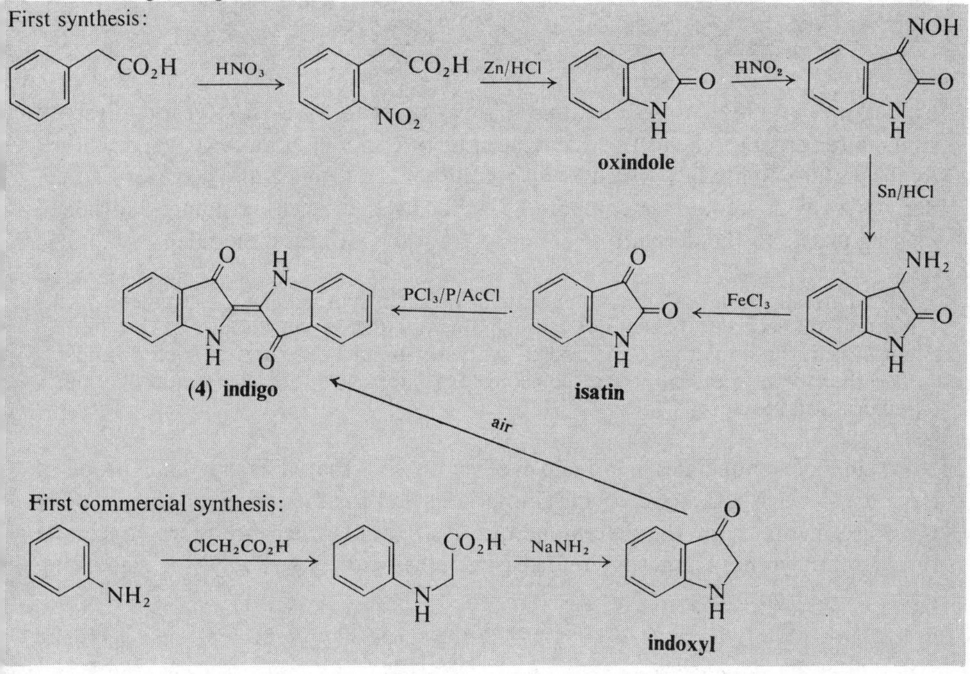

First synthesis:

oxindole

(4) indigo   isatin

First commercial synthesis:

indoxyl

This activity marked the beginning of a science-based, and particularly German, organic chemical industry. It ruined the cultivators of the natural product and also those manufacturers of coal-tar chemicals who unwisely looked for a quick profit in arbitrary and empirical recipes.

In a more academic field, we may note a fine example of the kind of synthesis which was possible by the end of the nineteenth century, that of α-terpineol (5)

(5)
α-terpineol

by Perkin[10] in 1904, using reactions known today to every first year University student of organic chemistry. The racemic α-terpineol produced was identical to the natural material, which is also commonly racemic, and gave crystalline derivatives which could be compared directly by their melting points and mixed melting points to the derivatives obtained from the natural material.

> '[This] investigation was undertaken with the object of synthesising . . . terpineol . . . , not only on account of the interest which always attaches to syntheses of this kind, but also in the hope that a method of synthesis might be devised of such a simple kind that there would no longer be room for doubt as to the constitution of [this] important [substance].'[10]

Perkin here enunciates a principle of synthesis—that it is the final proof of structure—which has stood chemists in good stead for the last sixty years. Today there are methods for the determination of structure which are more conclusive than a total synthesis, and different reasons must usually be found for embarking upon a total synthesis.

The syntheses described so far are all simple in one very important respect: the compounds are not optically active, nor do they contain more than one chiral centre.* This eases the problem, first (and trivially), because no resolution has to be done, secondly, because mixtures of diastereoisomers are not created in any one reaction. The importance of this latter point will become obvious in the later chapters of this book. It can be illustrated now by the chemistry of α-terpineol, the synthesis of which has just been described. Hydrogenation of α-terpineol (5)

and its mirror image   and its mirror image   and its mirror image
     (±)-(5)   (±)-(I)   (±)-(II)

gives a mixture of *two* diastereoisomeric products (I and II), which differ in the *relative* configuration of the methyl and the hydroxypropyl groups. Had we wished to synthesise one of these compounds, or make use of one of them in a subsequent synthesis, we would now have had to embark on a *separation* (in principle always possible but in practice not always easy), and in addition, we would have had to *identify which was which*. The additional problems these operations create can be considerable. As an alternative to the separation of I and II, we could have searched the literature for a reagent or a reaction which was stereoselective in favour of the one we want. Most likely, we would have to deduce, from a stereoselectivity observed with compounds of closely related structure, what reagent might give higher yields of the diastereoisomer we want. If no such reagent is available, then we would search for an alternative synthetic route, into which we might hope to incorporate the desired stereoselectivity. Examples of stereoselective reactions will be found later in this book, but the knowledge needed to plan a stereoselective synthesis was not available in the nineteenth or early twentieth century; stereoselective synthesis began to be feasible only in the 1940's.

Stereo*specificity* is sometimes distinguished from stereo*selectivity*.[11] The distinction is not always adhered to, but in some circumstances it is very useful and I have adopted it throughout this book. When one isomer of a pair of stereoisomers reacts to give more of one product (A) than another (B), and when the other isomer of the pair gives more of product (B) than of (A), then the reaction is said to be *stereospecific*. Thus the addition of bromine to maleic acid gives *meso*-dibromosuccinic acid, whereas the addition to fumaric acid gives *racemic*-dibromosuccinic

---

* Chirality (Greek $\chi\varepsilon\iota\rho$ = hand) means handedness. A chiral centre is thus a more precise description of what used to be called, rather loosely, an asymmetric centre.

acid; in another example, dimethyl fumarate and cyclopentadiene give one Diels-Alder adduct, with the methoxycarbonyl groups *trans* to each other, whereas dimethyl maleate gives the diastereoisomeric adduct with the methoxycarbonyl groups *cis* to each other. Both the addition reaction and the Diels-Alder reaction are therefore stereospecific. Stereo*selective* reactions, on the other hand, are *all* those reactions in which more of one stereoisomer is generated than of the other. Thus, while all stereospecific reactions are stereoselective, such reactions as the reduction of camphor to give predominantly isoborneol, or the hydrogenation of terpineol to give more of II than of I,[12] are merely stereoselective.

At this point it will be as well to establish the convention for the representation of stereochemistry which will be used throughout this book. Beyond this point, only one of the mirror image pair of a racemate will be drawn, and the text should make clear the point at which resolution, if carried out, removed the unwanted enantiomer. It is important to have clear in one's mind the distinction between *enantiomer* and *diastereoisomer*. For example, (+)-glucose (III) is the enantiomer of (−)-glucose (IV) but (+)-mannose (V) is a diastereoisomer of

(III)

III

(V)   (IV)

(+)-glucose (or (−)-glucose), differing in the configuration at one (or four) chiral centres. Separation of enantiomers is by resolution; separation of diastereoisomers, which have different chemical and physical properties, is done by such familiar techniques as fractional crystallisation, chromatography, and so forth.

At the stage in the history of synthesis which is our concern in this chapter, it was not usually possible to choose stereoselective reactions because the principles of conformation, of steric hindrance, and of the stereospecificity of many reactions were not known at that time. The problems created by the generation of a new chiral centre—problems of stereochemical control, of separation, and of identification—generally depended for their solution upon the favourable intervention of chance. An example where no such help was forthcoming is found in the synthesis of myoinositol (7) which was first claimed[13] to have been done by what seems to modern eyes the most inelegant of paths: the catalytic hydrogenation of hexa-hydroxybenzene (**6**). Although in this process eight different configurational isomers (one a racemic pair) of **7** could in principle have been produced, an 80% yield of **7** was claimed when a platinum catalyst was used. A more recent re-examination[14] discredits this claim and shows that, when Raney nickel is used as a catalyst, several stereoisomeric products, **7** among them, can indeed be produced.

(**6**) $\xrightarrow{H_2/cat.}$ (**7**) myoinositol + **other isomers**

This cautionary tale illustrates the difficulties which the presence of several chiral centres can present. It also shows, by contrast, how great was the achievement of Emil Fischer in the synthesis[15] of glucose and other sugars in 1890. It is not belittling Fischer's achievement to say that his good luck as well as his exceptional experimental skill must have played a large role in the synthesis: he showed that a product, known as acrose, which he obtained from acrolein dibromide (**8**) and alkali (and by several other similar routes), contained fructose (racemic of course) (**9**). While three chiral centres were set up in what must be a complex

(**8**) $\xrightarrow{Ba(OH)_2}$ (**9**) fructose + (**10**) sorbose + other hexoses

"acrose"

sequence of reactions leading to **9**, some factor evidently favoured the configurations found in fructose and sorbose (**10**). The fructose, with phenylhydrazine, gave racemic glucosazone (**11**). The remaining steps were a sequence of adjustments of the oxidation level and the configuration at the top two carbon atoms.

CH=NNHPh

"acrose"   1. PhNHNH₂   PhNHN= H   →   H₂O →   CHO
           2. Separate   H,,, OH   O= H
                              HO   OH   R OH
                                   H
                              OH   (±)-glucosone   R =   H,,, OH
                                                                        HO   OH
(11) (±)-glucosazone   H
                                                                            OH

Zn/HCl ↓

HO,,, CO₂H   HO   HO
   H H   ←Br₂/H₂O HO,,,   ←Na/Hg   O= H
      OH   H H   R OH
   R   R OH

(12) (±)-mannonic acid   **(±)-mannitol**   **(±)-fructose**

1. resolve
2. pyridine/140°

H,,, CO₂H   OH   OH   CHO
HO α H   ≡   HO₂C H   →   O H   Na/Hg   H,,, H
H,,, OH   HO   OH   HCl   OH   →   HO   H
HO   OH   H   OH   O   OH   HO   OH
     H   H   H   OH   H   OH
   OH   OH   OH   OH

(13) (−)-gluconic acid   **(+)-glucono-γ-lactone**   **(+)-glucose**

Fischer chose these reactions from his large experience of the reactions of sugars; they are remarkable chiefly for the reaction in which mannonic acid **(12)** was converted, by epimerisation at the carbon α to the carboxyl group, to the fortunately more stable diastereoisomer gluconic acid **(13)**. This reaction shows how, for successful synthesis, the configurational complexities of natural products must be appreciated and tamed. Had this step been omitted, mannose (V) would have been obtained instead of glucose. In this particular case the actual configurations were still unknown at the time of the synthesis; in the absence of the skill and techniques which a chemist today might hope to deploy in controlling the development of the chiral centres, a knowledge of what the actual configuration was would have been no help to Fischer. Of course that knowledge was itself worth having; and it was Fischer's next achievement to unravel the stereochemical relationships of the sugars. It is noteworthy that one of the steps in the sequence, the reduction of fructose to mannitol, is stereospecific. We are today able to predict that this kind of reaction would be stereospecific; it is a case where Cram's rule[16] (or the Karabatsos modification[17]) applies. It is also noteworthy that the resolution of the mannonic acid (12) was done with morphine; it so happens that morphine, when it was eventually synthesised (see p. 50), was resolved with tartaric acid, the resolution of which, in the hands of Pasteur, constituted the final step in completing a total synthesis without recourse to anything more 'natural' than a human being.

By the end of the nineteenth century, then, the nature of a total synthesis was appreciated and its achievement highly regarded. Of course the limited range of reactions available, the difficulty of stereochemical control and of separating mixtures, and the problems of determining the structures of reaction products restricted what could be done.

In order to provide a historical background, I have, so far, laid out these syntheses without detailed discussion. The reactions used were not always very general, nor are they especially instructive. In the remaining chapters I shall be more specific about individual steps. Since I have chosen a more or less chronological order, the increasing resources and understanding of the exponents of synthesis will become apparent. From now on, however, I shall be far more concerned with the chemical reactions themselves than with their historical significance.

References

1. F. Wöhler, *Annalen der Physik und Chemie*, **12**, 253(1828); the work was important enough to be published in French, *Ann. Chim. (France)*, **37**, 330, and English, *Quart J. Sci.*, Pt 1, 491 (1828), as well as the original German.
2. A brief account of this matter is given in H. M. Leicester, *Historical Background of Chemistry*, Wiley, New York, 1956. For a reassessment of the significance of Wöhler's work see, D. McKie, *Nature*, **153**, 608 (1944).
3. H. Kolbe, *Annalen*, **54**, 145 (1845), where the word synthesis is first used, *ibid.*, 186.
4. M. Berthelot, *Chimie organique fondée sur la synthèse*, Paris 1860, translated in H. M. Leicester and H. S. Klickstein, *A Source Book in Chemistry 1400–1900*, McGraw-Hill, New York, 1952 and Harvard University Press, Cambridge, 1965, p. 430.
5. A. M. Butlerow, *Jahresbericht uber die Fortschritte der Chemie*, 475 (1863) and 496 (1864).
6. W. H. Perkin, *J. Chem. Soc.*, 232 (1862); British Patent No. 1984 (1856); Perkin was 18 at the time of his discovery. The structures were worked out by O. Fischer thirty years later.
7. First synthesis, C. Graebe and C. Liebermann, *Ber.*, **2**, 332 (1869); first commercial synthesis, C. Graebe, C. Liebermann and H. Caro, *Ber.*, **3**, 359 (1870) and W. H. Perkin, *J. Chem. Soc.*, 133 (1870).
8. A. Baeyer, *Ber.*, **11**, 1296 (1878).
9. K. Heumann, *Ber.*, **23**, 3431 (1890).
10. W. H. Perkin, jun., *J. Chem. Soc.*, 654 (1904).

11. H. E. Zimmerman, L. Singer and B. S. Thyagarajan, *J. Amer. Chem. Soc.*, **81**, 108 (1959), footnote 16; E. L. Eliel, *Stereochemistry of Carbon Compounds*, McGraw-Hill, New York, 1962, p. 436.
12. R. H. Eastman and R. A. Quinn, *J. Amer. Chem. Soc.*, **82**, 4249 (1960); H. van Bekkum, D. Medema P. E. Verkade and B. M. Wepster, *Rec. Trav. chim.*, **81**, 269 (1962).
13. H. Wieland and R. S. Wishart, *Ber.*, **47**, 2082 (1914).
14. R. C. Anderson and E. S. Wallis, *J. Amer. Chem. Soc.*, **70**, 2931 (1948).
15. E. Fischer, *Ber.*, **23**, 799 (1890).
16. D. J. Cram and D. R. Wilson, *J. Amer. Chem. Soc.*, **85**, 1245 (1963); Cram's rule of asymmetric induction is mentioned in a number of textbooks, see for example: J. March, *Advanced Organic Chemistry*, McGraw-Hill, New York, 1968, p. 90; J. D. Roberts and M. C. Caserio, *Basic Principles of Organic Chemistry*, Benjamin, New York, 1964, p. 600; E. S. Gould, *Mechanism and Structure in Organic Chemistry*, Holt, New York, 1959, p. 549; E. L. Eliel, *Stereochemistry of Carbon Compounds*, McGraw-Hill, New York, 1962, p. 68.
17. G. J. Karabatsos, *J. Amer. Chem. Soc.*, **89**, 1367 (1967).

Tropinone (2) and Cocaine (3)

Atropine, the racemate of structure **1**, is an abundant alkaloid with marked physiological properties, the best known of which is a capacity to cause the pupil of the eye to dilate. A key degradation product obtained during the determination of its structure by Willstätter is tropinone (**2**). Willstätter synthesised tropinone in a long sequence of reactions[1] involving the elimination of amino groups by Hofmann exhaustive methylation, the addition of bromine to double bonds, the substitution of bromide by dimethylamine, the elimination of hydrogen bromide by dimethylamine and so on. These steps are illustrated in the sequence on p. 18. The yields are mostly very good. A reinvestigation by modern methods would probably show that many of the products were mixtures, though in many cases it would not matter. However, in spite of the good yields, the arithmetic demon ensures, in a long synthesis, that the overall yield will be low—in this case, 0·75 per cent based on cycloheptanone. It is instructive to see how quickly even 90 per cent, multiplied by itself a few times, drops (see p. 101). The effect of a poorer step or two is heartbreaking in practice, particularly if these steps come near the middle. When this happens, either many of the early steps have to be done on a cumbersomely large scale or many of the subsequent steps have to be worked out on a distressingly small scale.

It is perhaps unfair to the genius of Willstätter to draw attention to this very important and general principle in the context of one of his outstanding achievements. It is done, at least partly, in order to accentuate the beauty and simplicity

(2)

of Robinson's synthesis of tropinone in one step.[2] This followed from a simple dislocation of the molecule:

Robinson described his thinking and his course of action thus:[2]

'By imaginary hydrolysis at the points indicated by the dotted lines, the substance may be resolved into succindialdehyde, methylamine and acetone.

... It was proved that tropinone is obtained in small yield by the condensation of succindialdehyde with acetone and methylamine in aqueous solution. An improvement followed on the replacement of acetone by a salt [calcium] of acetone dicarboxylic acid. The initial product is a salt of tropinone dicarboxylic acid, and this loses two molecules of carbon dioxide with the formation of tropinone when the solution is acidified and heated'.

Although the kind of reaction involved in this synthesis was not unknown at the time, it was not by any means so familiar as its 'name-reaction' status makes it today. It is, of course, a double Mannich reaction. The methylamine and the aldehyde combine reversibly to give an appreciable concentration of a protonated imine (I). The enol of acetone dicarboxylic acid (II) was also present in solution. The combination of this enol, nucleophilic on carbon, and the electrophilic protonated imine resulted in carbon–carbon bond formation.

This process, the creation of *carbon–carbon* bonds, is a central preoccupation in the design of most of the important syntheses we shall be discussing. The problem was avoided in Willstätter's synthesis, because he chose a starting material which already possessed all the carbon–carbon bonds of tropinone and therefore required no more than modification of the functional groups, the introduction and removal of heteroatoms, and so forth.

The remaining steps in Robinson's synthesis, after the formation of the first carbon–carbon bond, were a repetition of the same sequence on the other side of the molecule (III → IV → V). The precise details of this reaction, both in the first and in the second carbon–carbon bond forming steps, are not known, and there are other plausible mechanisms than that given here.

(III)   (IV)   (V)

Since both stages of this process are *ring forming* reactions, and especially since five- and six-membered rings are being formed, these later stages are likely to be relatively fast. The intermolecular component of the first stages—the formation of I and the combination of I and II—makes this kind of reaction slower than the otherwise very similar but intramolecular reactions—the cyclisation of III to IV and of IV to V. The five- and six-membered rings are formed with such great ease because the frequency of collision between the reacting termini in such unstrained systems is much greater than between two separate molecules. The easy formation of five- and six-membered rings will be a feature of many syntheses in this book. Smaller, more strained, rings and larger rings present special problems and usually call for special reactions. A good example, in which both a four-membered and a nine-membered ring have to be set up, is discussed in the section on caryophylene, p. 137.

The family relationship of the Mannich reaction to the aldol condensation and the Perkin reaction should be obvious. These reactions represent one of the simplest and most used ways of forming carbon–carbon bonds—the combination of an enol, or enolate, as a nucleophile and a carbonyl group as electrophile. The great majority of the reactions which establish carbon–carbon bonds involve the combination of nucleophilic carbon with electrophilic carbon. The homolytic processes—radical addition and radical combination—and the concerted processes—Diels-Alder reactions and carbene insertions—are, of course, important and will make their appearance in due course.

Having described Robinson's synthesis, we must now bring the story further up to date. In extensive researches,[3] Schöpf and his co-workers raised the yield to more than 90 per cent, principally by a judicious choice of buffer solution.

Other dialdehydes may be used: glutardialdehyde (4), for example, gave[4] pseudopelletierine (5), an alkaloid in the bark of the pomegranate.

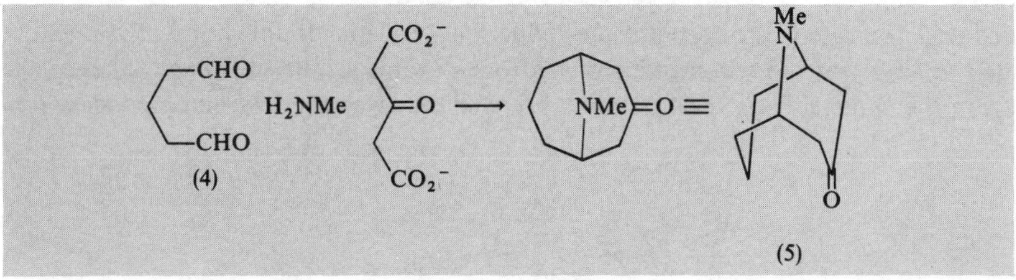

Willstätter also found[5] a very simple route to tropinone:

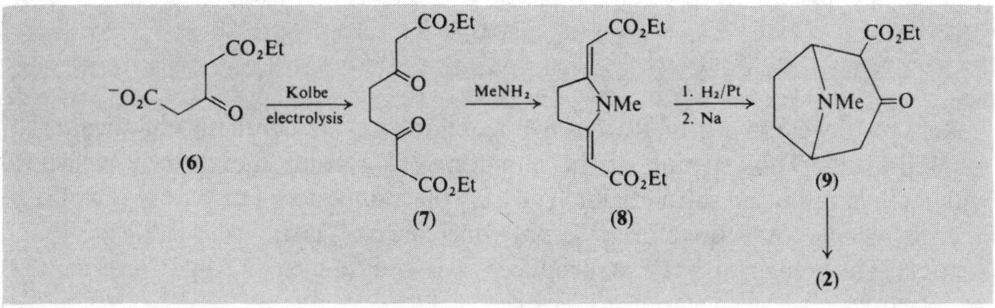

In this synthesis one carbon–carbon bond ($6 \rightarrow 7$) was set up in an electrolytic and, very likely, homolytic step and the other (tetrahydro-**8** = VI $\rightarrow$ **9**) by a Dieckman condensation (the cyclic version of the Claisen condensation). This latter reaction was mechanistically closely related to the aldol process, being in the first step (VII, arrows) the combination of an enolate with a carbonyl group. The

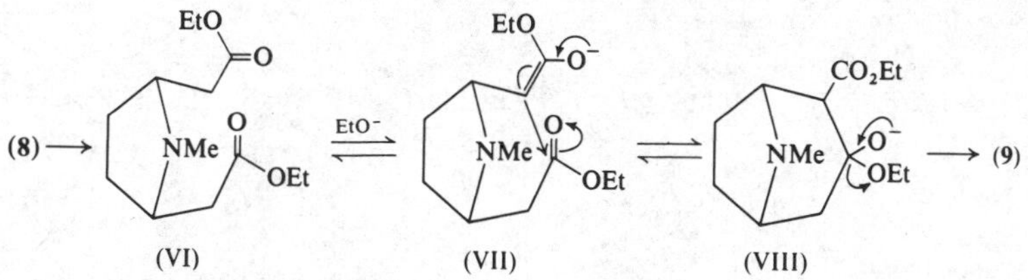

intermediate (VIII) resulting from this carbon–carbon bond forming process had, as a natural consequence of the different oxidation level of the (ester) starting material, a different conclusion, namely ejection of the ethoxide ion (VIII, arrows).

The Dieckman condensation worked because both ester groups were on the same side of the 5-membered ring in VI. To modern eyes, this is what would be expected, because the hydrogens were introduced by a hydrogenation step: hydrogenations usually result in *cis* delivery of hydrogen to the less hindered side of a double bond. In this case, we do not know whether each double bond was reduced separately or not, but if the reduction went one double bond at a time, the delivery (IX) of the second pair of hydrogen atoms would still have taken place from the same side as the first pair, because that was the less hindered side.

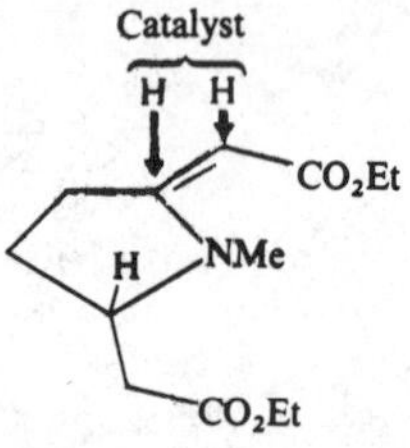

(IX)

Another development of Robinson's reaction was the synthesis[6] of cocaine (**3**) by Willstätter. The reaction of the monoester of acetone dicarboxylic acid (**10**) with methylamine and succindialdehyde gave the monoester of tropinone dicarboxylic acid, which could therefore lose only one carboxyl group on acidification and heating. The ketoester (**11**) so produced was reduced to the hydroxyester (**12**)

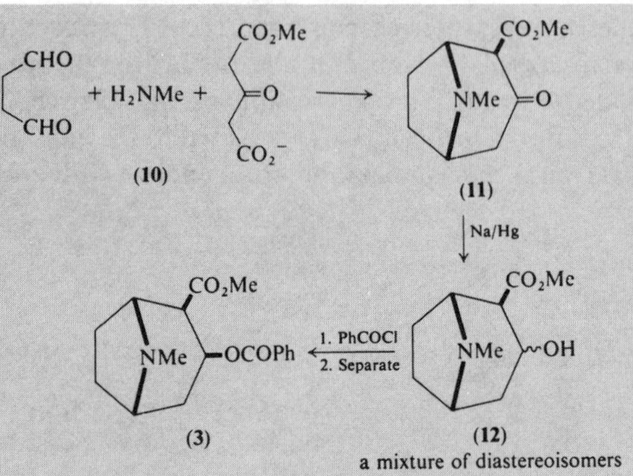

(10)　　　　(11)

(3)　　　　(12)
a mixture of diastereoisomers

with sodium amalgam, and at this point stereochemical problems appeared for the first time. There were four possible racemic pairs of hydroxy esters (**12**) only two of which were actually produced; fortunately, the methoxycarbonyl group had taken up the natural configuration (*cis* to the NMe bridge) in **11** (and retained it in **12**). This kind of good luck is quite frequent in the synthesis of natural products, where the most stable isomer is often the desired one. Actually the relative and absolute configurations of the various chiral centres in atropine and cocaine were not known at the time this synthetic work was being done; I have been able to put these details in the drawings in this chapter because of more recent work.[7] However, in this synthesis it was not necessary to know what the actual configuration was: benzoylation of the mixture of isomers (12) and fractional crystallisation to remove the unwanted isomer gave racemic cocaine, which was then resolved to give material identical to the natural product (3).

Robinson's work on tropinone was influential in another respect. He was quick to draw attention[8] to the significance of the near-physiological conditions used in the reaction, and also to the close relationship of the starting materials to compounds known to occur in nature. He argued that tropinone, and many other alkaloids, could be derived in nature by processes very similar to those which he showed took place in the laboratory. The principle of biogenetically patterned synthesis has been a stimulating one and has inspired many syntheses.[9]

Question

In the first Willstätter synthesis p. 18 many of the reactions could have given mixtures, either of isomers or of compounds of different structure. Test your general sense of the likelihood of these events by considering what could have gone wrong at each step.

References

1. R. Willstätter, *Ber.*, **34**, 129 and 3163 (1901); **29**, 936 (1896). For an account in English see H. L. Holmes in *The Alkaloids*, R. H. F. Manske and H. L. Holmes (Ed.), Vol. 1 p. 288–292, Academic Press, New York, 1950.
2. R. Robinson, *J. Chem. Soc.*, 762 (1917).
3. C. Schöpf, G. Lehmann and W. Arnold, *Angew, Chem.*, **50**, 783 (1937).
4. R. C. Menzies and R. Robinson, *J. Chem. Soc.*, 2163 (1924).
5. R. Willstätter and M. Pfannenstiel, *Annalen*, **422**, 1 (1921); R. Willstätter and M. Bommer, *ibid.*, 15; see also W. Parker, R. A. Raphael and D. I. Wilkinson, *J. Chem. Soc.*, 2433 (1959).
6. R. Willstätter, O. Wolfes and H. Mäder, *Annalen*, **434**, 111 (1923).
7. G. Fodor in *The Alkaloids*, R. H. F. Manske (Ed.), Vol. 6, p. 145–159, Academic Press, New York, 1960.
8. R. Robinson, *J. Chem. Soc.*, 876 (1917).
9. E. E. van Tamelen, *Biogenetic-type Syntheses*, in *Progress in the Chemistry of Organic Natural Products*, **19**, Springer Verlag, Vienna, 1961.

Cyclooctatetraene

Cyclooctatetraene is not a natural product, but belongs rather to a class of organic compounds which is worthy of synthesis because of its interest to the theory of organic chemistry. In the present case the question needing an answer was whether cyclooctatetraene would, like benzene, be unusually stable; would it undergo substitution rather than addition reactions: would it, in short, be aromatic? The compound was first synthesised by Willstätter in 1911,[1] as follows:

This route closely parallels the one he first used in the synthesis of tropinone (p. 18). The starting material, pseudopelletierine, is an alkaloid the structure of which Willstätter had worked out and which was therefore available to him. The product, he found,[1,2] was not at all like benzene; and there the matter rested for over thirty years.

In 1947, Cope and Overberger[3] repeated this synthesis, using now synthetic pseudopelletierine (see p. 21). They did so because there was by that time some doubt[4] about the structure of Willstätter's product, which, it was thought, could have been styrene. They confirmed Willstätter's work and showed that the product was not styrene. They also showed that their product was identical to cyclooctatetraene prepared by a method developed, during the war, at BASF in Germany by Reppe. Reppe's work led to high yields of cyclooctatetraene by the catalysed tetramerisation of acetylene.

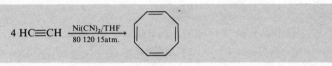

The work was eventually published,[5] but at that time was circulating in American government reports of German technological developments. Today the BASF plant is the only source of cyclooctatetraene; gifts of this material have been used all over the world as starting material for countless syntheses of other compounds of theoretical interest, such as cyclobutadiene (see p. 144), bullvalene, [16]annulene, and the homocycloheptatrienyl cation.[6]

Questions

The Reppe synthesis probably involves coordination of the nickel(II) ion with four molecules of acetylene and two of cyanide. Draw structures for this intermediate which look to be reasonable precursors of cyclooctatetraene. If you are ambitious, consider whether the cyclotetramerisation in the complex is 'allowed' by the Woodward-Hoffmann rules. For some discussion of the theory see reference 7.

References

1. R. Willstätter and E. Waser, *Ber.*, **44**, 3423 (1911).
2. R. Willstätter and M. Heidelberger, *Ber.*, **46**, 517 (1913).
3. A. C. Cope and C. G. Overberger, *J. Amer. Chem. Soc.*, **69**, 976 (1947); **70**, 1433 (1948).
4. For a review of these doubts and for an account of the state of the concept of aromaticity at this time see J. W. Baker, *J. Chem. Soc.*, 258 (1945).
5. W. Reppe, O. Schlichting, K. Klager and T. Toepel, *Annalen*, **560**, 1 (1948).
6. For review of cyclooctatetraene chemistry see G. Schröder *Cyclooctatetraen*, Verlag Chemie, Weinheim, 1965.
7. G. N. Schrauzer, P. Glockner and S. Eichler, *Angew. Chem. Internat. Edn.*, **3**, 851 (1964).

Callistephin Chloride

(1)   (2)

Callistephin chloride (1) is a red flower pigment. It was first isolated from the purple-red aster, and was crystallised in 1916.[1] The structure was established by Robinson in the course of his studies of the anthocyanin pigments. Part of the proof of structure was by synthesis. The degradative work had established that the pigment was a glucoside of the anthocyanidin pellargonidin (2), in which the glucose was attached to the 3- or 7-hydroxyl group. By attaching the glucose at an early stage onto the hydroxyl group which would become the 3-substituent, Robinson was able to prove that this was the site of substitution in the natural pigment and, at the same time, to accomplish the first synthesis of a flower pigment.[2] The synthesis, especially the reaction[3] which established the benzpyryllium ring system, took advantage of Robinson's earlier extensive work on the synthesis of the aglucones, such as pellargonidin (2).

Anisole was acylated in a Friedel-Crafts reaction, using chloroacetyl chloride and aluminium chloride, to give mainly the *p*-isomer (3) (see p. 28). We should

note that this reaction established a new carbon–carbon bond, that the Friedel-Crafts reaction was a particularly useful way of doing this with aromatic compounds, and that in outline the mechanism was one of attack by the aromatic ring—activated in its capacity as a nucleophile by the methoxy group—on the electrophilic acyl cation, thus establishing the new bond; this process was followed by loss of the proton to restore aromaticity.

We should be particularly conscious that we have here a powerful electrophile reacting with a comparatively poor nucleophile whereas the aldol type reaction has a powerful nucleophile (the enolate ion) attacking a comparatively poor electrophile, the carbonyl group. The combination of the relatively weakly nucleophilic anisole and the relatively weakly electrophilic carbonyl group would not (without strong acid catalysis) lead to reaction. The combination of a powerful nucleophile, a carbanion, and a powerful electrophile, a carbonium ion, would no doubt be extremely fast were it possible, in the ordinary course of events, to get both species into the same reaction mixture.

We have established, in the paragraph above and in the discussion of the Mannich reaction on p. 19 that carbon–carbon bond formation is usually the combination of an electrophilic carbon species with a nucleophilic carbon species, and we have established that the formation of carbon–carbon bonds is the most important constructive task in any organic synthesis. Having done this now, at an early stage, we shall be able henceforth to look only at mechanistic points of special rather than general interest. It is implicit that you should understand each reaction in broad mechanistic terms before proceeding to the next. Some of the more interesting ones are discussed specifically; some are left as questions at the ends of chapters. Throughout, you should check that every reaction which is used, but not discussed, is either understandable or already familiar. The accumulated experience of knowing just which nucleophilic species will react with just which electrophilic species is the basis of a working knowledge of organic chemistry.

The chloro substituent in **3** was next displaced by acetate ion in an SN2 reaction, the adjacent carbonyl group making this process unusually easy, for reasons which are not entirely clear. The solvent, acetic anhydride, acetylated the phenolic group. Alkaline hydrolysis gave the diol (**4**).

It was next necessary to attach the glucose to the hydroxyl group (labelled ω on **4**) destined to become the oxygen function on C-3 of the final product (**1**). However, there were two hydroxyl groups in **4** and it was necessary to prevent the 4-hydroxyl group from becoming bound to the glucose. This was prevented with a *protecting group*, a much used device. Such a device must have the property of masking a functional group while reaction is carried out; it must be capable of surviving the reaction conditions, and of being released by a suitably chosen reagent at the appropriate time. There is a great variety of protecting groups available for most situations;[4] many of them will be found in subsequent sections of this book.

In this particular case the protecting group used was the acetyl group. It was attached to the phenolic hydroxyl group selectively because the phenolate ion, produced in alkali, was much more nucleophilic than the ω-hydroxyl group, which was not extensively ionised in alkali. Now that the phenolic hydroxyl group was protected, the sugar could be attached to the ω-hydroxyl group. The sugar itself was protected as its tetraacetyl derivative, and made more reactive as its α-glucosidyl bromide. In the presence of silver ions the bromide ion could be displaced even by such poor nucleophiles as the ω-hydroxyl group, and the product (5) was then that of inversion at the glucosidic link. This fragment was then joined to 6 in the reaction pioneered for this particular kind of purpose—the establishment of the flavylium ring system.[3] Phloroglucinaldehyde was protected by mono-benzoylation to give 6, and the two components were mixed in ether with dry hydrogen chloride. The product (**7**) crystallised as red plates from the mixture. The reaction, was, in essence, an acid-catalysed aldol process establishing the new carbon–carbon bond. The subsequent formation of the heterocyclic ring, which was six-membered and aromatic, is not surprising: the new bonds were formed in an entirely straightforward way.

The protecting groups were removed by alkaline hydrolysis, which also, in all probability, opened the pyryllium ring. However, reacidification restored the ring and the red colour. After extensive purification, pure material was isolated, identical to the natural product.

Questions

1. Chloroacetyl chloride could, in principle, have reacted, with displacement of either chlorine, to give acylation or alkylation. Why was acylation observed?
2. The attentive and critical reader may have noticed that, in the course of the Friedel-Crafts reaction the ether function was cleaved. How did the reagents involved achieve this helpful short cut?
3. The detailed steps in the acid catalysed condensation of **6** with **5** are not, of course, known. Examine the possible pathways in this reaction and decide which is the most likely course of events.

References

1. R. Willstätter and C. L. Burdick, *Annalen*, **412**, 149 (1916).
2. A. Robertson and R. Robinson, *J. Chem. Soc.*, 1460 (1928).
3. D. D. Pratt and R. Robinson, *J. Chem. Soc.*, 1577 (1922).
4. For a review of protecting groups see J. F. W. McOmie, *Advances in Organic Chemistry, Methods and Results*, Vol. 3, Interscience, New York, 1963.

Thyroxine

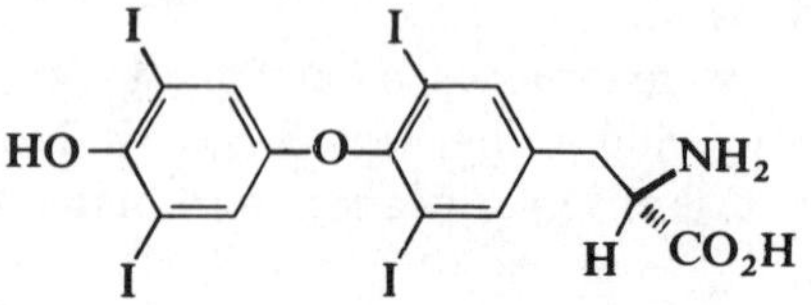

(1)

The hormone thyroxine (**1**) furnishes another example in which the synthesis[1] (by Harington and Barger in 1927) played a vital part in the proof of a structure which had been, up to that time, tentative at best.

The synthesis began with two aromatic nucleophilic substitutions: first of a diazonium group by iodide and secondly of the iodide by the *p*-methoxyphenate ion. The first of these reactions probably involved a rate-determining loss of nitrogen to give a very reactive cation, which rapidly captured iodide ion. The second however followed a different course—addition followed by elimination.

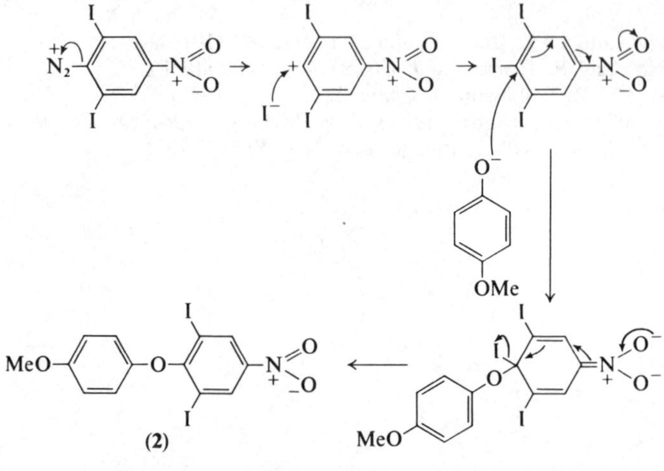

(2)

Reduction of the nitro group of **2**, diazotisation of the amino group, a Sandmeyer reaction with cupric cyanide, and Stephen reduction gave the aldehyde (**3**) in a conventional sequence for the modification of aromatic substituents. Condensation of the aldehyde with hippuric acid gave the azlactone (**4**) in an Erlenmeyer reaction.

Hydrolysis (or rather ethanolysis) and reduction with phosphorus and hydriodic acid gave the amino acid (5) in which the protecting group had also been cleaved off. Of the two rings, that containing the free hydroxyl group was much the more nucleophilic in alkaline solution. Accordingly, iodination, with iodine and potassium iodide in aqueous ammonia, gave thyroxine (1) in which the obviously active *o*-positions had become substituted. The synthetic material was identical with natural thyroxine, which at that stage had been extracted only in racemic form. Subsequently the synthetic material was resolved,[2] and the natural material was isolated in optically active form;[3] later still, the true natural material was shown[4] to be in the L-series of amino acids. As a result of the known, or reliably inferred, positions of the iodine atoms in the synthetic material, the position of the iodine atoms in thyroxine was established and the structure was therefore proved.

Straightforward though this synthesis was, it was not particularly good for the preparation of the large quantities of thyroxine needed for medical use. One of many syntheses which has actually been used on an industrial scale is that[5] due to Hems of Glaxo laboratories.

L-Tyrosine (6) was used as a starting material, and all steps were, as a result of painstaking investigation, chosen to avoid subsequent racemisation. Indeed the specific choice of the protecting groups was made on the basis of their safety in this respect. Again, the diphenyl ether link was made by aromatic nucleophilic

substitution; but this time *p*-toluenesulphonate ion was displaced; this leaving group is, of course, a very common one, being conveniently prepared from hydroxyl groups, which are not themselves good leaving groups. Subsequent steps, shown in the sequence (8 → 1), are noteworthy chiefly for the care taken to get conditions which would give good yields. The overall yield was 26 per cent, which represents an average of 86 per cent over the nine stages.

Question

The starting materials, 2,6-diiodo-4-nitroaniline, hydroquinone monomethyl ether, and tyrosine may be prepared by many routes. Devise simple syntheses of these compounds.

References

1. C. R. Harington and G. Barger, *Biochem. J.*, **21**, 169 (1927); for a small improvement in the conversion of **4** to **5**, see C. R. Harington and W. McCartney, *ibid.*, 852.
2. C. R. Harington, *Biochem. J.* **22**, 1429 (1928).
3. C. R. Harington and W. T. Salter, *Biochem. J.*, **24**, 457 (1930).
4. A. Canzanelli, C. R. Harington and S. S. Randall, *Biochem. J.*, **28**, 68 (1934).
5. J. R. Chalmers, G. T. Dickson, J. Elks and B. A. Hems, *J. Chem. Soc.*, 3424 (1949).

Ascorbic Acid

(1)

Ascorbic acid was first isolated from orange juice and crystallised in 1928, and was later identified as the antiscorbutic factor of lemon juice. Its structure **(1)** was elucidated by Haworth, and two syntheses were reported in 1933.[1,2] Since the configuration at C-5 is the S-configuration, ascorbic acid belongs to the unusual L-series of sugars. One way to achieve a synthesis of ascorbic acid is to start with a readily available natural (and therefore D-) sugar, in which case the easiest way to solve the problem of configuration is to invert the two ends of the starting material so that C-1 becomes a CH_2OH group and C-6 a group at a higher oxidation level. If an L-sugar derivative is to be the result of this operation, the configuration at C-2 must be like that in glucose and galactose, but not like that in, say, mannose.

Glucose   Galactose   Mannose

Thus, of the first syntheses, that due to Haworth and Hirst began with D-galactose, and that due to Reichstein began with D-glucose. They are both, in fact, *partial* syntheses; of course, glucose and galactose had been synthesised, so that conversions of them to ascorbic acid constituted total syntheses in principle, if not in practice. The synthesis by Haworth and Hirst[1] is described below.

D-Galactose **(2)** was protected as the di-isopropylidene derivative **(3)**. This protecting group is well known to be formed readily with *cis*-vicinal diols; the result is that only the hydroxyl group on C-6 is exposed to the next reagent. Oxidation, therefore, gave the acid **(4)**, from which the protecting groups were removed to give **5**. Sodium amalgam reduced the (masked) aldehyde group to give galactonic acid (**6**). The carboxylic acid group was derived from C-6, and the original C-1 now bears a single hydroxyl group; thus the 'ends' of the sugar have been turned over: the galactonic acid (**6**) is in the desired L-series.

The amide of the galactonic acid (**6**) was treated with sodium hypochlorite to give the L-lyxose (**7**). This reaction is known as a Weerman degradation: it follows the pathway—*N*-chloroamide (I) to isocyanate (II)—of the Hofmann degradation of amides, but the next step leads directly to the aldehyde (III = 7) and not, by hydrolysis of the isocyanate, to the amine, as in the Hofmann degradation.

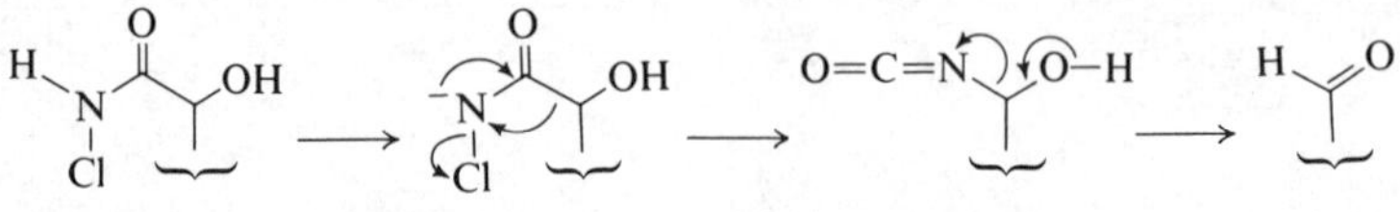

(I)   (II)   (III)

The aldehyde (7) (no doubt masked* as the hemiacetal) was oxidised to L-xylosone (8) by forming the osazone and then removing the phenylhydrazine. Addition of hydrogen cyanide and hydrolysis of the cyanohydrin (9) gave L-ascorbic acid (1). Lactonisation of the intermediate (IV) under the acidic conditions, was followed by enolisation of the resultant α-ketolactone (V). (α-Dicarbonyl

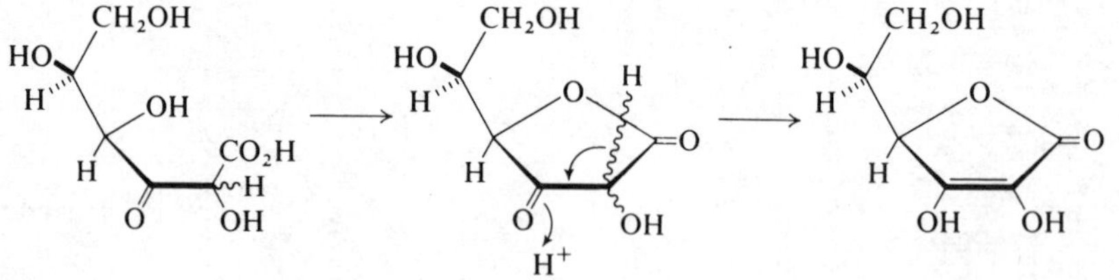

(IV)   (V)   (VI)

compounds, in cyclic systems, enolise with great readiness and enolisation generally goes to completion, as was the case in this synthesis: the enediol system (VI = 1) is the stable tautomeric form of ascorbic acid.) The sequence from an aldehyde to an α-hydroxy acid used in this synthesis is called the Kiliani ascent of a series.

The first syntheses, such as that above, were not suitable for large scale production. An early industrial synthesis of ascorbic acid starts from D-glucose. Actually this industrial synthesis, using microbiological oxidation, is not a total synthesis, even in principle; the use of microorganisms, which have not themselves been synthesised, disqualifies the route as a total synthesis in the usually accepted academic

* The aldehyde group of 7 is very likely to be involved in a tautomeric equilibrium:

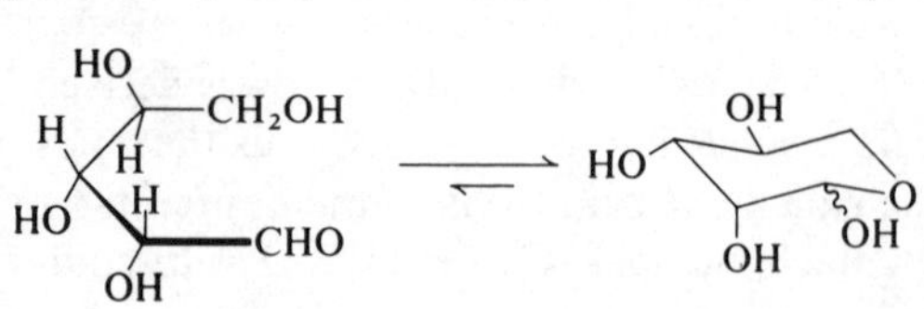

but, even if the cyclic tautomer is predominant, there will still be some free aldehyde present under the reaction conditions, and any free aldehyde that reacts will be replaced as equilibrium is again set up. Thus the cyclic tautomer is said to have a *masked* aldehyde group.

sense. From the point of view of industrial practice, this was simply the best way to do the reaction. It is largely based on work of Reichstein.[3]

Glucose (10) was reduced to sorbitol (11), and the sorbitol oxidised, using the bacterium *Acetobacter suboxydans*,[4] to L-sorbose (12). Bacteria and fungi are very widely used in industrial synthesis; we shall meet another example in the Velluz steroid synthesis (p. 130). They have the inestimable advantage over the usual reagents of being able, in many cases, to carry out highly specific reactions, whether in a polyfunctional molecule or in a completely unfunctionalised one. The problem is to find the species and the strain which will do what is wanted and no more. Such a search may be laborious, and is usually only worthwhile when the product is needed on an industrial scale. They are not convenient laboratory reagents, although some enzymes, such as ribonuclease, which have rather more predictable properties, certainly are. In this case, once again, the ends have been turned over and we are now in the L-series, to which ascorbic acid belongs.

The sorbose (12) was protected as the di-isopropylidene derivative (13), which was then oxidised, at the only free functional site, to the protected acid (14). Hydrolysis gave 2-oxogulonic acid (15), which was esterified and the ester (16) then treated with base. In the ensuing reaction, the C-4 hydroxy anion (VII)

(VII)   (VIII)

displaced the ester group to give a lactone (VIII) similar to that in the earlier synthesis. The lactone again enolised to give the stable tautomer of ascorbic acid (1).

Questions

1. The ready enolisation of β-diketones (IX) is well known. Equilibrium often lies well over to the right in cyclic systems, but much less so in acyclic systems. The same observation holds for α-diketones (X) such as the case of V and VIII above. Why should cyclic systems be more completely enolised at equilibrium than are acyclic systems?

(IX)   (X)

2. And why should cyclic α-diketones enolise almost completely, when cyclic ketones are almost completely non-enolised at equilibrium?

References

1. R. G. Ault, D. K. Baird, H. C. Carrington, W. N. Haworth, R. Herbert, E. L. Hirst, E. G. V. Percival, F. Smith and M. Stacey, *J. Chem. Soc.*, 1419 (1933).
2. T. Reichstein, A. Grüssner and R. Oppenauer, *Helv. Chim. Acta*, **16**, 1019 (1933).
3. T. Reichstein and A. Grüssner, *Helv. Chim. Acta*, **17**, 311 (1934).
4. P. A. Wells, J. J. Stubbs, L. B. Lockwood and E. T. Roe, *Ind. Eng. Chem.*, **29**, 1385 (1937).

Mesoporphyrin-IX (1) and Haemin (2)

(1)   (2)

Haemin is the protein-free pigment responsible for the colour of blood and is intimately involved in oxygen transport. It belongs to the general class of compounds known as porphyrins. As a result of extensive degradative work, Hans Fischer established the structure (2). The structural work involved the synthesis of many degradation products, including a number of porphyrins; this use of synthesis in structure determination was particularly necessary in order to establish the order of the substituents around the perimeter. As an example of how synthesis may be used in structure determination, we shall discuss first the synthesis of one of these degradation products, mesopophyrin-IX (1).

Oxidation of haemin had given various substituted maleimides, and reduction had given various substituted pyrroles. From the structure of these fragments, from a knowledge of the molecular formula, and from a knowledge of other reactions which haemin underwent, it had been suggested[1] that haemin had a macro-cyclic structure in which the perimeter was substituted with four methyl, two vinyl, and two propionic acid residues, with one methyl group in each ring (I). There are fifteen possible arrangements within these limits; so, in order to simplify the problem, Fischer degraded the side chains. Reduction of the vinyl groups of haemin and decarboxylation gave a tetraethyl-tetramethyl porphyrin (II). There were now only four possible structures. These four, known as aetioporphyrins, were synthesised, but they turned out to be indistinguishable from one another, and it was therefore not possible to assign a structure to the aetioporphyrin derived from haemin. It thus became necessary to synthesise each of the fifteen meso-porphyrins (III) obtained by reducing the vinyl groups and removing the iron. Of these, the one called mesoporphyrin-IX proved to be identical to the meso-porphyrin derived from haemin. The synthesis of this mesoporphyrin[2] is illustrative of the methods used in the synthesis of all of the mesoporphyrins and the aetioporphyrins.

(I)*   (II)   (III)*

Usually the first part of the overall strategy in porphyrin synthesis is the prepara-tion of the four pyrroles, which become rings A, B, C, and D. One pair of pyrroles (3) and (4), destined to become rings A and B, was prepared by conventional routes which took advantage of standard pyrrole syntheses and of the well known nucleo-philic reactivity of pyrroles, greater in the 2-position than in the 3-position.

The synthesis of the pyrrole (8), destined to become both rings C and D, began with a straightforward pyrrole synthesis and continued with a selective hydrolysis and decarboxylation of 5 to give 6. This reaction began with the attack of water on a protonated carbonyl group (IV or V, solid arrows). In this step, the delocalisa-tion (IV and V, dotted arrows) must be interrupted; the water evidently attacked as in V because delocalisation to a 3-substituent (V, dotted arrows) is less effective

(IV)   (V)

* In the drawings (I and III), the groups placed at the side of the molecule are those known to be present, but whose disposition in rings A, B, C and D was not known.

in stabilising the pyrrole than is delocalisation to the 2-substituent (IV, dotted arrows). (This explanation is essentially the same as that used to account for the greater nucleophilicity of pyrroles in the 2-position.) Thus it is the 3-ethoxycarbonyl group which is most easily expelled.

The remaining steps were straightforward until the bromination of **7** to give **8**. The detailed mechanism of this step is not, of course, known and several pathways could reasonably be offered. One possibility is the sequence (VI → VII).

The next step in the overall strategy was to join the pyrroles in pairs. The two pyrroles (3) and (4) were joined by treatment with hydrobromic acid. The mechanism for the combination of 3 and 4 is nicely illustrative of pyrrole reactivity.

The protonated aldehyde (IX from 4) was attacked by the free pyrrole (VIII = 3), and the subsequent steps then led to the salt (X = 9), a member of a remarkably stable class of compounds, the dipyrrylmethene salts; at least part of this stability came from the good delocalisation (X, arrows). The dipyrrylmethene salt (9) was brominated to give the dipyrrylmethene salt (10), presumably by a mechanism similar to the bromination step (VI → VII).

The pyrrole (8) was coupled with itself to give the dipyrrylmethane (11), a surprising reaction which probably took the course (XI → XII), in which formaldehyde is the other product. The dipyrrylmethanes as a class are rather unstable,

unlike the dipyrrylmethenes; but in this case the ethoxycarbonyl groups had a stabilising influence on the pyrrole rings as a result of overlap like that shown in IV. Alkaline hydrolysis of 11 gave the acid (12) which, with bromine, was brominated, decarboxylated and oxidised to the dipyrrylmethene salt (13). The brominative decarboxylation step (12 → 13) is reminiscent of the bromination of salicylic acid to give tribromophenol, and took the course, shown in outline, (XIII → XIV).

The reactions involved in the syntheses of the two dipyrrylmethene salts (10) and (13) illustrate many important aspects of pyrrole chemistry. Particularly important in the context of porphyrin synthesis are the ease of bromination of side chains and the special ease with which the bromide can then be displaced. The final step in the overall strategy of porphyrin synthesis is the combination of the two dipyrrylmethene salts in a reaction which again takes advantage of this side-chain reactivity. It is done, characteristically, by what has been described as a bold and brutal method, that of fusing the two salts in succinic acid at 180–190°, in the air, for one to two hours. The reaction only rarely gives good yields of porphyrins, this particular example being one of them: **10** and **13** give meso-porphyrin-IX (**1**) in 31 per cent yield. In outline the mechanism may involve the loss of a proton from **10** to give the nucleophilic system of XV, which can then, by addition and elimination, displace bromide from **13**. Since **13** is symmetrical, there is no ambiguity about which end is attacked first, and thus only one (porphyrin) product is formed in this case. Repetition of the process can establish a similar bond on the other side of the ring (XVI); at some stage (e.g. XVII) loss of bromine must take place to give the porphyrin ring.

* The double-headed arrow used in 13 and XVI implies that there was an intermediate. Thus, in a simpler case, saponification can be written either in full:

(XVIII)

or it can be abbreviated:

The double-headed arrow has saved us having to draw the intermediate (XVIII).

There are many variations of this sequence which could be offered as alternative mechanisms. The one above is presented as one of the simplest accounts possible. The establishment of a sixteen-membered ring calls for some comment. Conditions of high dilution are not used, although in normal alicyclic systems of this size they would be necessary. The rigid planar chromophores of the two halves of XVI greatly reduce the number of bonds about which rotation will occur. The carbon atoms which become bonded in the next step are thus not too infrequently within bonding distance, and reaction proceeds without a large and unfavourable entropy-of-activation term. Evidence that the low yields normally encountered in such porphyrin syntheses are not due to the large size of the ring, but rather to the large number of other pathways which polyfunctional molecules like **10** and **13** can follow, is to be found in the many, more recent and much improved porphyrin syntheses.[3] Proper control of the functionality of the reacting molecules and careful choice of reagents and conditions have led to several methods which give excellent yields by any standards.

With the synthesis of mesoporphyrin-IX (and of the other fourteen mesoporphyrins) completed, it was possible for Fischer, with some conviction, to assign the structure (2) to the blood pigment itself. He then embarked upon the synthesis[4] of that molecule, using the methods well worked out in the structural investigation one small part of which was described above.

Once again he prepared two dipyrrylmethene salts, one of which (**16**) was new, but could readily be prepared by the standard procedure of mixing a pyrrole having a free 2-position (**14**) with a pyrrole 2-aldehyde (**15**). The other dipyrrylmethene salt (**13**) was the same CD component used in the synthesis of mesoporphyrin-IX. The fusion in succinic acid was more normal in giving **17** in very low yield, less than 2 per cent. In this case, the presence of free 3-positions on the pyrrole rings may well have contributed to deflect the starting materials towards unwanted products. The formation of the ring system, in the step analogous to the cyclisation of XVI, is followed by an oxidation step in which, overall, two atoms of hydrogen are removed. The fact that in the synthesis of mesoporphyrin-IX a reductive step (the debromination XVII) is necessary, and that in the haem synthesis it is an oxidative step, shows that it is possible in Fischer's method of porphyrin synthesis to show a fine disregard for the oxidation level of the starting materials. In any normal synthesis it is not advisable to expect such reductions or oxidations to come to one's aid, and very considerable care is normally taken to check that oxidation levels are right. In this case the stability of the porphyrin ring system must have contributed to the ease with which the hydrogen or bromine atoms are expelled.

Porphyrins behave in many respects like aromatic rings: they can be regarded as substituted diaza[18]annulenes. They possess a paramagnetic ring current, as judged in the n.m.r. spectrum by the unusually low-field position of resonance of the meso hydrogens, marked $\alpha - \delta$ on **17**. They also show a strong tendency to give porphyrin products rather than addition products; that is, they readily undergo electrophilic substitution reactions.

The remaining steps of the synthesis of haemin involve the introduction of the vinyl groups at the free 2- and 4-positions of the porphyrin (**17**), using such an electrophilic substitution in the first step. Friedel-Crafts reaction on the iron complex gave, after removal of the iron with acid, the diacetyl derivative (**18**). The ketone groups were reduced with an unconventional but not unheard-of reagent, hot potassium hydroxide in ethanol, which presumably delivered hydride ion in a manner analogous to the delivery of the hydride ion in the Meerwein-Ponndorf reduction and in the Cannizzaro reaction.

The diol corresponding to **18** was dehydrated by heating, either alone under vacuum or with acid. Addition of ferric ion gave the natural product (**2**).

In many ways this work represents, principally by the number of steps involved, one of the highest achievements in synthesis before the Second World War. However, the war marked a turning point in the history of synthesis: afterwards, new techniques, both of structure determination (the spectroscopic methods and X-ray crystallography) and of separation (chromatography), were introduced. These techniques, and the greatly increased knowledge of reagents and of mechanisms which intense investigation has yielded, have combined to increase the number and complexity, expecially stereochemical complexity, of the molecules which have fallen to the attack of chemists intent upon synthesising them. Woodward's synthesis of chlorophyll, see p. 112, when compared with Fischer's of haemin, is an appropriate example.

Questions

1. The pyrrole (**5**) was prepared in 'a straightforward pyrrole synthesis'. Write out a mechanistic scheme for this synthesis.
2. The final step of the haemin synthesis (before the introduction of the iron) was the rather easy dehydration of the diol. Why was this diol so easy to dehydrate?

References

1. W. Küster, *Z. physiol. Chem.*, **82**, 463 (1912).
2. H. Fischer and G. Stangler, *Annalen*, **459**, 53 (1927).
3. For reviews of more recent porphyrin syntheses see R. L. N. Harris, A. W. Johnson and I. T. Kay, *Quart. Rev.*, **20**, 211 (1966); A. H. Jackson and G. W. Kenner, *Nature*, **215**, 1126 (1967).
4. H. Fischer and A. Kirstahler, *Annalen*, **466**, 178 (1928); H. Fischer and K. Zeile, *Annalen*, **468**, 98 (1929).

Morphine

(1)

Morphine (1) presents a much greater synthetic challenge than the compounds we have met already. Here, we have, for the first time, a bridged structure with several chiral centres and several different functional groups juxtaposed. The only earlier synthesis where problems of this kind and difficulty were met and solved was that of quinine.[1] We shall now see how these problems were dealt with in the case of morphine.

The early stages[2] of the morphine synthesis took advantage of a stable platform, the naphthalene ring, on which to build up a set of functional groups. The mono-benzoate of 2,6-dihydroxynaphthalene was nitrosated in the α-position next to the free hydroxyl group, as a result of the usual and well known capacity of β-naphthols to be nucleophilic in the α-position, a phenomenon ascribed to 'bond fixation' in the older resonance terminology. Reduction of the nitroso group and immediate oxidation gave the quinone (2), which was reduced, and was protected for the next few stages by a conventional methylation. Alkaline hydrolysis removed the other protecting group, and the nitrosation-reduction-oxidation sequence was repeated to give the quinone (3).

The anion of cyanoacetic ester was then used to attack the 4-position of the quinone, in a Michael-type of reaction, and, by incorporating ferricyanide in the mixture, the intermediate hydroquinone was oxidised to the quinone (4). This was hydrolysed and then readily decarboxylated, being both an α-cyanoester and a vinylogous β-ketoester.

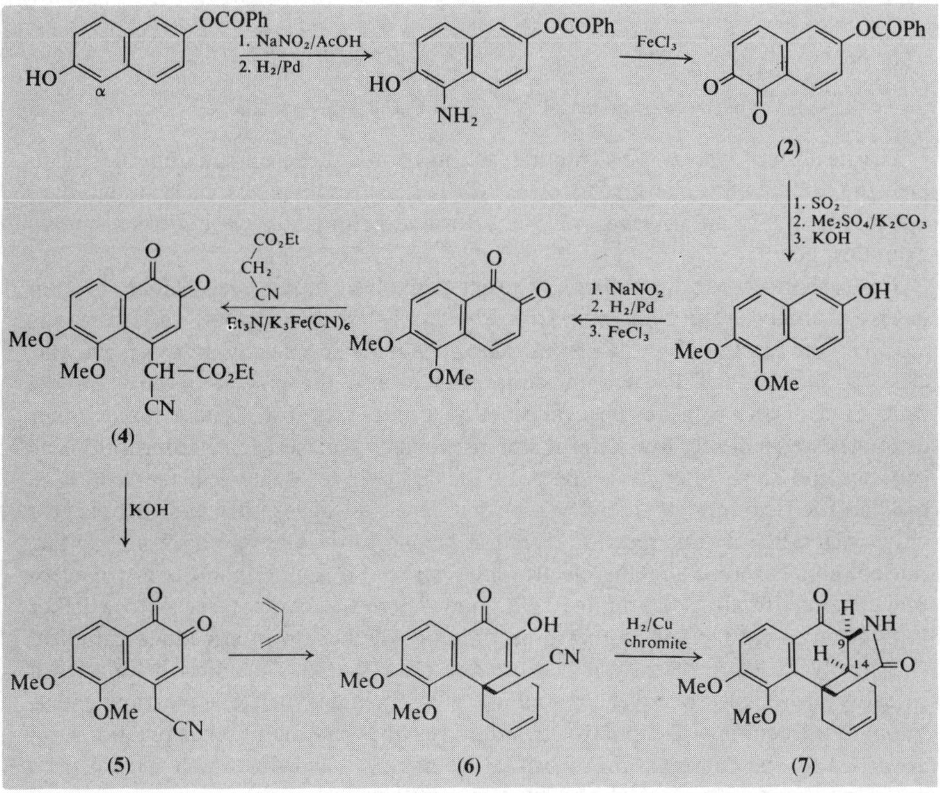

(2)

(4)

(5)   (6)   (7)

The word vinylogous calls for some justification.[4] Whenever a functional group such as a carbonyl group is conjugated with a double bond, its functionality is often relayed through that double bond. Thus a conjugated carbonyl compound is often attacked by nucleophiles at the conjugate position, as in the Michael reaction, for example:

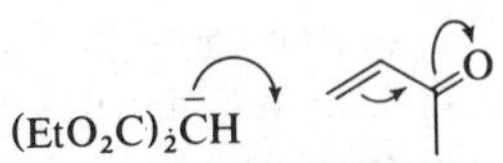

$(EtO_2C)_2CH$

Similarly, when two functional groups, which normally have an influence on each others' properties when they are adjacent, are instead separated by a double bond, that influence persists, and they are now described as being the vinylogous equivalent of the more simple arrangement. An amide, for example, is stabilised by overlap, and so is the vinylogous amide system:

amide delocalisation   vinylogous amide delocalisation

The next step[3] was a Diels–Alder reaction of butadiene on the quinone (5) to give the adduct (6). The first-formed 1,2-diketone enolised, as such systems are apt to do, to give an α-hydroxy-α,β-unsaturated ketone, a grouping often known as a diosphenol.

The carbon–carbon bond-forming process involved in the Diels–Alder reaction is exceptionally useful and interesting. Not only are two carbon–carbon bonds set up in one reaction, but the process also goes stereospecifically. In this particular case the stereospecificity was not important, since the proton initially on the same side of the ring as the cyanomethyl group was lost. But in many other examples we shall see how helpful stereospecificity can be in providing only one product and that, often, the one with the desired stereochemical features. The mechanistic features of the Diels–Alder reaction have intrigued people for many years. Plainly this reaction does not belong to the class we have met so far, nucleophilic carbon attacking electrophilic carbon: a double bond is not nucleophilic enough to attack a conjugated ketone. There was controversy as to whether both bonds were formed at the same time or whether an intermediate diradical was involved. More recently it has become clear that this reaction belongs to a category, now called pericyclic reactions, which includes the Cope rearrangement, 1,5-hydrogen shifts, 1,3-dipolar cycloadditions and very many other reactions, all governed by the conservation of orbital symmetry. The rules which govern pericyclic reactions were worked out by Woodward and Hoffmann[5] who, stimulated by observations first made in the course of synthetic work directed towards vitamin B_{12},[6] went on to demonstrate spectacularly how much new insight can come from a chance observation made in the course of a quite different enterprise.

The next stage in the synthesis is one which does not belong in any list of well categorised reactions—catalytic hydrogenation of 6 gave the lactam (7). Indeed, this reaction was at first thought to have taken a different course.[7] Only when infrared spectroscopy—a newly available tool—revealed the absence of the nitrile function was it obvious that a lactam had been formed in this step.[8] This was fortunate, because the longer sequence originally envisaged became unnecessary: most of the morphine skeleton was now present.

In practice the hydrogenation reaction ($6 \rightarrow 7$) had been developed with the closely similar but much more simply prepared model compounds (I) and (II), in which the methoxy groups are not present. This use of *model compounds* is typical of many large syntheses in which new processes are to be used. In most cases model reactions, if chosen well, work with the real thing, as was the case in the synthesis of morphine. But some sorry disappointments have been experienced, when apparently quite minor changes of structure have led to behaviour very different from that of the model.

(I)   (II)

With most of the morphine skeleton in hand, some development of functionality was needed next. The first steps in this direction involved the removal of unwanted carbonyl groups: Wolff–Kishner reduction removed the ketone and lithium aluminium hydride the lactam. The N-methyl group was incorporated either after the reduction, or, better, by alkylation of the intermediate lactam.

Although the Wolff–Kishner reduction was done under strongly basic conditions, and there was a chiral centre adjacent to the ketone group, no change in configuration was expected, because the configuration at this centre was the only one possible in this ring system. This property of the smaller rings greatly simplified the synthesis. Although three chiral centres were present in 7 and 8, only one mirror-image pair was actually obtained: one centre, C-14, had been set up the way it was because the catalyst had delivered hydrogen to the less hindered side of 6, the side opposite the cyanomethyl group; the second centre, C-9, was the only possible arrangement in which the new ring could be formed.

The amine (8) was now resolved, using the salt with dibenzoyltartaric acid. This procedure gave an opportunity for direct comparison of the synthetic material with material obtained by degradation of the natural product and thus to provide reassurance that both materials had the expected structures. Also it was possible to use, for all subsequent steps in the synthesis, material actually obtained, less arduously, from natural sources. A compound used in this way is usually referred to as a *relay*. It is, of course, essential to be especially thorough in proving that the synthetic sample of the relay has the same structure as that of the sample obtained by degradation. Although the complete synthesis has not, in practice, been carried out, the *idea*, indeed the certainty, that it can be carried out quite properly constitutes a synthesis.

The next step was to hydrate the isolated double bond using dilute acid. Fortunately, hydration introduced the hydroxyl group in the 6-position, and for good reasons. Protonation of the double bond was followed by pseudoaxial attack

(7) → 1. KOH/N₂H₄ 2. NaH/MeI 3. LiAlH₄ → (8) → 1. resolved 2. H₂SO₄/H₂O → (9)

(9) → PhN⁺Me₃ ŌH → (10) → BuᵗO⁻/Ph₂CO → (11) → Br₂ → (12)

(12) → 2,4-DNPH → (13) → HCl/H₂O/Me₂CO → (14) → 1. H₂/Pt 2. 2Br₂ 3. 2,4-DNPH → (15)

(15) → HCl/H₂O/M → (16) → LiAlH₄ → (17) → Py.HCl → (1)

by the nucleophile, water. If this attack took place at the 6-position, the water could be delivered from the side bearing the protonated ethylamine bridge, which, because of its charge, had attracted the water molecules (III). This is not an entirely

(III)   (IV)

convincing argument and could hardly have been predicted with any confidence. The alternative axial mode of attack, resulting in a C-7 hydroxyl, would have had the advantage of having no axial substituent in the way of the nucleophile (IV). That the desired C-6 hydroxylated product was actually obtained was confirmed by the subsequent steps.

The methyl ether at C-4 of **9** was then selectively cleaved, somewhat unusually, under alkaline conditions. This unexpected reaction was developed because the Wolff–Kishner reduction, used earlier, had been observed to give some demethylation at this position. The 6-hydroxyl group of **10** was oxidised, using a variation of the Oppenauer method to give the ketone (**11**). Bromination of **11** gave a dibromo derivative (**12**) which, with 2,4-dinitrophenylhydrazine, gave, as usual with this reagent and α-bromoketones, the dehydrobrominated derivative (**13**). The yield was low, no doubt because bromination did not go exclusively to C-7. The dinitrophenylhydrazone actually isolated had the inverted configuration at C-14. The configuration in **13** was known, from studies on other morphine derivatives, to be the more stable; but only when the conjugated double bond of **13** had been introduced was there a pathway—loss of proton by 'enolisation' and reprotonation—available for epimerisation. The reaction conditions were acidic enough for the process to have taken place at the first opportunity.

When syntheses like this one are being designed, the chemist must not only face the problem of setting up chiral centres with the desired configuration, but must also examine all the steps in the sequence for the possibility that epimerisation may take place at these centres. Alternatively, as in the present case, if he has set up such a centre in the wrong sense, he must later arrange suitable functionality in order to permit epimerisation, and he must ensure, at the same time, that the stereochemical features are such as to make inversion energetically profitable. We shall see other examples of such thinking later in this book, most notably in the Woodward syntheses of steroids and reserpine.

The hydrazone (13) was hydrolysed to a ketone (14). Reduction of the double bond of this ketone, rebromination with two equivalents of bromine and treatment with 2,4-dinitrophenylhydrazine gave the derivative (15), presumably by way of a 5,7-dibromo derivative with intramolecular displacement of the C-5 bromide. Hydrolysis gave the ketone (16), from which the bromine was removed by hydrogenolysis, using, somewhat unusually, lithium aluminium hydride, which also reduces the ketone group to give the alcohol (17) stereoselectively and in the right sense.

The conversion of 17, which is codeine, to morphine (1) was known to be possible with pyridine hydrochloride, a reagent much used for ether cleavage when acidic conditions must, as in this case, be avoided. This completed the synthesis of morphine.

Questions

1. In the reductive cyclisation (6 → 7) the conditions were 130°/4 hr, 27 atmospheres of H_2, in alcohol with copper chromite catalyst. Consider the possible mechanisms of this process.
2. The relative stabilities of 14 and its C-14 epimer are given in the text. Would you have been able to predict from conformational considerations that 14 would be the more stable?
3. The reaction of phenylhydrazines with α-bromoketones to give the phenylhydrazone of the α,β-unsaturated ketone requires some explanation, since phenylhydrazine is not a strong enough base, under these conditions, to have caused dehydrobromination of the bromoketone as the first step.

References

1. R. B. Woodward and W. E. Doering, *J. Amer. Chem. Soc.*, **67**, 860 (1945); the synthesis of quinine is also described at length in a useful introductory textbook by R. E. Ireland, *Organic Synthesis*, Prentice-Hall, Englewood Cliffs, New Jersey, 1969, pp. 123–139.
2. M. Gates *J. Amer. Chem. Soc.*, **72**, 228 (1950).
3. M. Gates and G. Tschudi, *J. Amer. Chem. Soc.*, **78**, 1380 (1956).
4. R. C. Fuson, *Chem. Rev.*, **16**, 1 (1935).
5. R. B. Woodward and R. Hoffmann, *Angew. Chem. Internat. Edn.*, **8**, 781 (1969).
6. R. B. Woodward in *Aromaticity*, Special Publication No. 21 of the Chemical Society, London, 1967.
7. M. Gates and W. F. Newhall, *J. Amer. Chem. Soc.*, **70**, 2261 (1948).
8. M. Gates, R. B. Woodward, W. F. Newhall and R. Kunzli, *J. Amer. Chem. Soc.*, **72**, 1141 (1950).

Cholesterol and Cortisone

(1)   (2)

The synthesis of steroids has been an active field and continues to be so. The products are of considerable pharmacological interest and are much used in medicine. The early synthesis of partially aromatic steroids—of equilenin in 1939[1] and oestrone in 1948[2]—set the stage for the first syntheses of the sterols proper which were completed in 1951.[3,4] Particularly important at that time for medicinal purposes was a new group of steroids, the corticoids, possessing 11-oxygenated functions, of which cortisone (2) is the key member. The first two syntheses of the sterols were completed almost simultaneously. The route described here is Woodward's;[3] it has the special advantage of incorporating functionality in ring c, from which the 11-oxo group of cortisone itself could be developed.

From the outset the problem of stereochemical control was an important factor in the design. Cholesterol (1) has eight chiral centres, numbered on 1, and therefore 256 possible isomers, very few of which are excluded by the nature of the ring system; the situation is thus unlike that discussed in the case of morphine. In a *bridged* ring system like that of morphine, the configuration at one bridgehead automatically determines the configuration at the other. This is not the case in *fused* ring systems like that of the steroids.

The first problem tackled was the CD ring system which was *trans* fused, and therefore had the less stable arrangement for a *six*-membered ring fused to a *five*-membered ring. A Diels–Alder reaction between 3 and butadiene gave the adduct (4) with, as a result of the concertedness of the cycloaddition, the *cis* ring fusion. However, when two *six*-membered rings are fused, the *trans* fusion is usually the more stable. In this case the trigonal centres and the substituents give 5 little advantage over 4, but a fortunate observation led to a simple way of converting 4 to 5. As a result of enolisation (towards C-14),* 4 was acidic enough to dissolve

* Steroid numbering is used throughout.

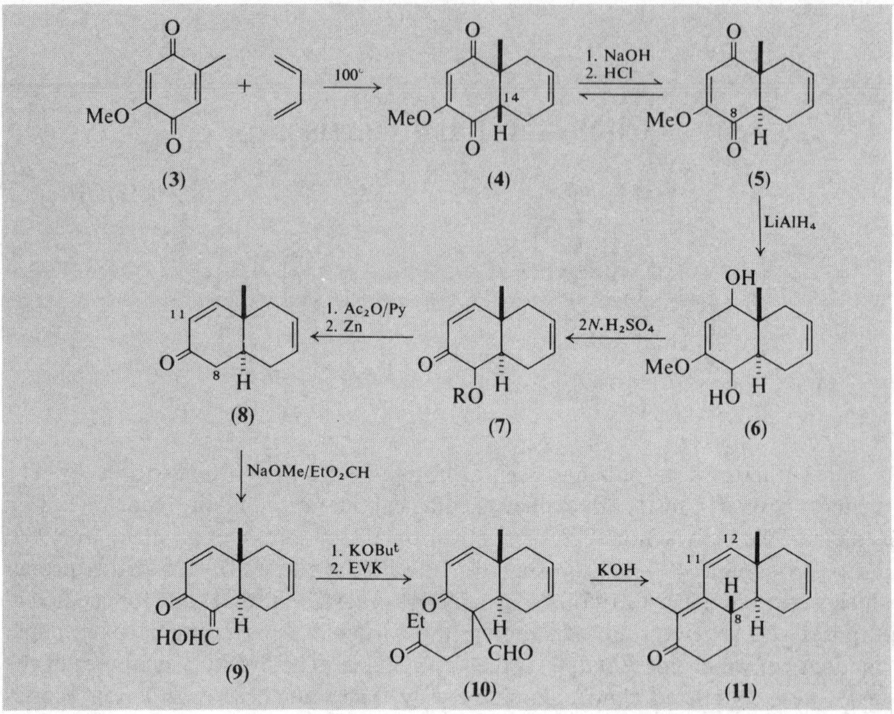

in aqueous alkali. Acidification of the solution of the enolate ordinarily gave a mixture of **4** and **5**; however, if seeds of **5** were deliberately added first, **5** could be made to crystallise out, without **4** making an appearance. The *trans* ring junction was now set up; the next steps removed the ketone group at C-8, which could otherwise lead it to return to a *cis* ring junction. Lithium aluminium hydride reduction gave the diol (**6**), the enol ether grouping of which was hydrolysed in aqueous acid. The resulting ketone, being a β-hydroxyketone, spontaneously lost water to give the unsaturated ketone (**7**, R = H). Reduction of the acetate of this ketone (**7**, R = Ac) with zinc gave the α,β-unsaturated ketone (**8**). This last reaction can be represented:

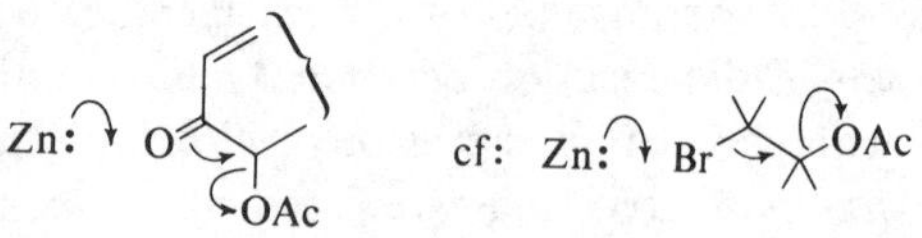

and was successful in the absence of acid. (The α,β-unsaturated ketone grouping is susceptible to metal reduction if the carbonyl group is protonated.) The unsaturated ketone (**8**) could enolise only towards C-8 which was therefore, a potentially nucleophilic carbon. The ketone was condensed with ethyl formate to give the hydroxymethylene derivative (9), in a Claisen condensation:

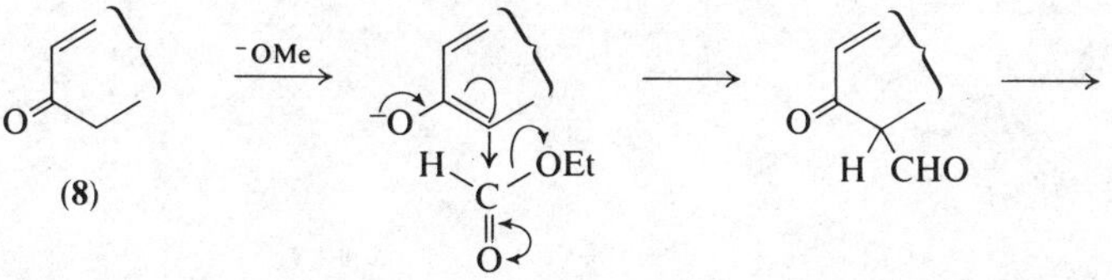

in which the product was completely enolised. The anion of 9 was readily generated with potassium t-butoxide, and then combined with ethyl vinyl ketone in a Michael reaction to give the adduct (10). When this was treated with potassium hydroxide, the formyl group was eliminated (the reverse of a Claisen condensation) and the diketone cyclised (in an aldol-type condensation) to give a single ketone (11).

This sequence (8 → 11) is a version of the *Robinson ring extension*, a reaction sequence designed expressly for the synthesis of steroids and used, naturally, in Robinson's own synthesis.[4]

Because the ketone (11) can enolise towards C-8, this centre can be equilibrated. Since it had been set up under equilibrating conditions, it should, therefore have had the more stable configuration. The configuration shown in 11 was expected to be the more stable on the grounds (i) that diaxial interactions across ring C were greater in the compound epimeric at C-8 and (ii) that the coplanarity of the conjugated system was better preserved in 11 than in its diastereoisomer.

At this stage the double bond between C-11 and C-12 had served its turn in preventing enolisation at the keto group in 8 towards C-11. Before removing it by hydrogenation however, the isolated double bond, even more easily reducible, had to be protected. Reaction of 11 with osmium tetroxide gave the diol (12) (the α-orientation being presumed), which was further protected as the acetonide (13) by reaction with acetone. Oxidising agents, like osmium tetroxide, are electrophilic, and it is not surprising, therefore, that the isolated double bond should have reacted rather than one of the double bonds conjugated to the ketone group. The isolated double bond was also, from a steric point of view, less deeply embedded. Hydrogenation of 13 gave 14, the disubstituted double bond being reduced much more readily than the tetrasubstituted double bond.

To make ring A, the potentially nucleophilic carbon C-10 had to be joined to the carbon atoms which would complete this ring. Now enolisation of 14 could make C-6, C-8, C-10 or C-11 nucleophilic, depending on the direction of enolisation. With ethyl formate and sodium methoxide, condensation took place at C-6 to give the corresponding hydroxymethylene derivative (15, R = OH) Reaction of this with methylaniline gave the vinylogous amide system of 15 (R = NMePh). This

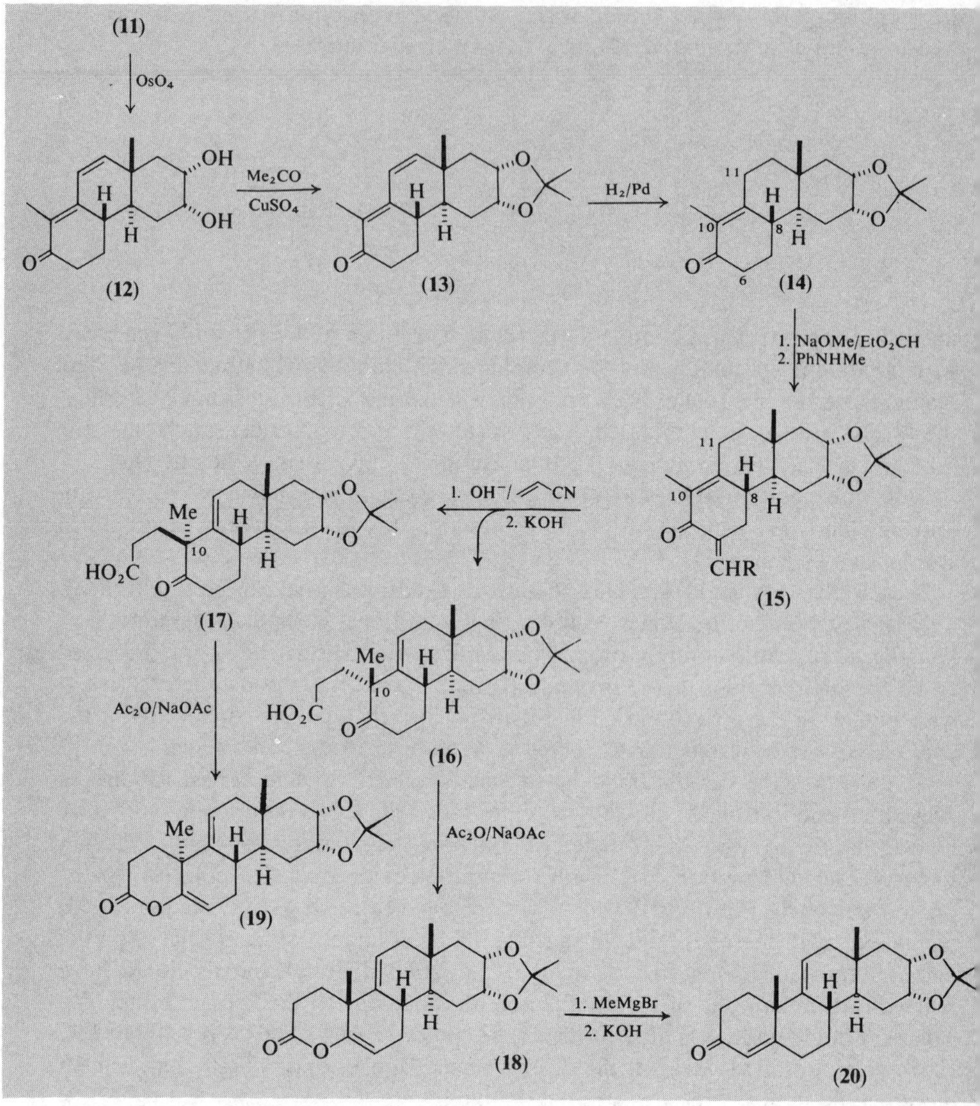

(11)

OsO₄

(12) $\xrightarrow[\text{CuSO}_4]{\text{Me}_2\text{CO}}$ (13) $\xrightarrow{\text{H}_2/\text{Pd}}$ (14)

1. NaOMe/EtO₂CH
2. PhNHMe

(15)

(17) $\xleftarrow[\text{2. KOH}]{\text{1. OH}^-/\text{CN}}$ (16)

Ac₂O/NaOAc

(19)

Ac₂O/NaOAc

(18) $\xrightarrow[\text{2. KOH}]{\text{1. MeMgBr}}$ (20)

protected derivative could now enolise only towards C-8 or C-11, in either case making C-10 nucleophilic. It often happens that in an extended enolate ion, such as I, the more substituted carbon, in spite of steric hindrance, is nevertheless the more nucleophilic (I, solid arrow); it is also usual for the first carbon—C-10 in this case—to be more nucleophilic than the second (C-11), even when the degree of substitution is the same. In the event, 15 with hydroxide ion (Triton B,

PhCH$_2$N$^+$Me$_3^-$OH, was used) and acrylonitrile gave, after hydrolysis of the nitrile group and the protecting group, the two products (**16** and **17**) of the Michael reaction at C-10. The two products differed only in the configuration at C-10, and

(I)

there was no α,β-unsaturated ketone (which would have resulted from attack at C-11 in I, dotted arrows). The ketoacids were readily separated, and both keto-acids gave enol lactones (**18** and **19**) with acetic anhydride and a trace of sodium acetate.

Of course at this stage in the synthesis it was not known which was which, and indeed it was not *known* that the isolated double bond was between C-9 and C-11 (rather than between C-8 and C-9). The subsequent steps cleared up both these matters just as an n.m.r. spectrum would have done, had this technique been available at the time. One of the lactones reacted smoothly with methyl magnesium bromide to give a diketone, which was cyclised by base (in an aldol reaction). The

other lactone was troublesome in this sequence. This sequence had been worked out earlier in the cholesterol series by other workers and was known to be an effective procedure for introducing the one-carbon unit of C-4. It was therefore likely that the minor enol lactone, which was the one behaving well and giving good yields, was the desired one, namely **18**. Thus the unsaturated ketone most easily obtained was probably **20**. Since the further elaboration of this compound did lead to the natural products, its structure was eventually confirmed.

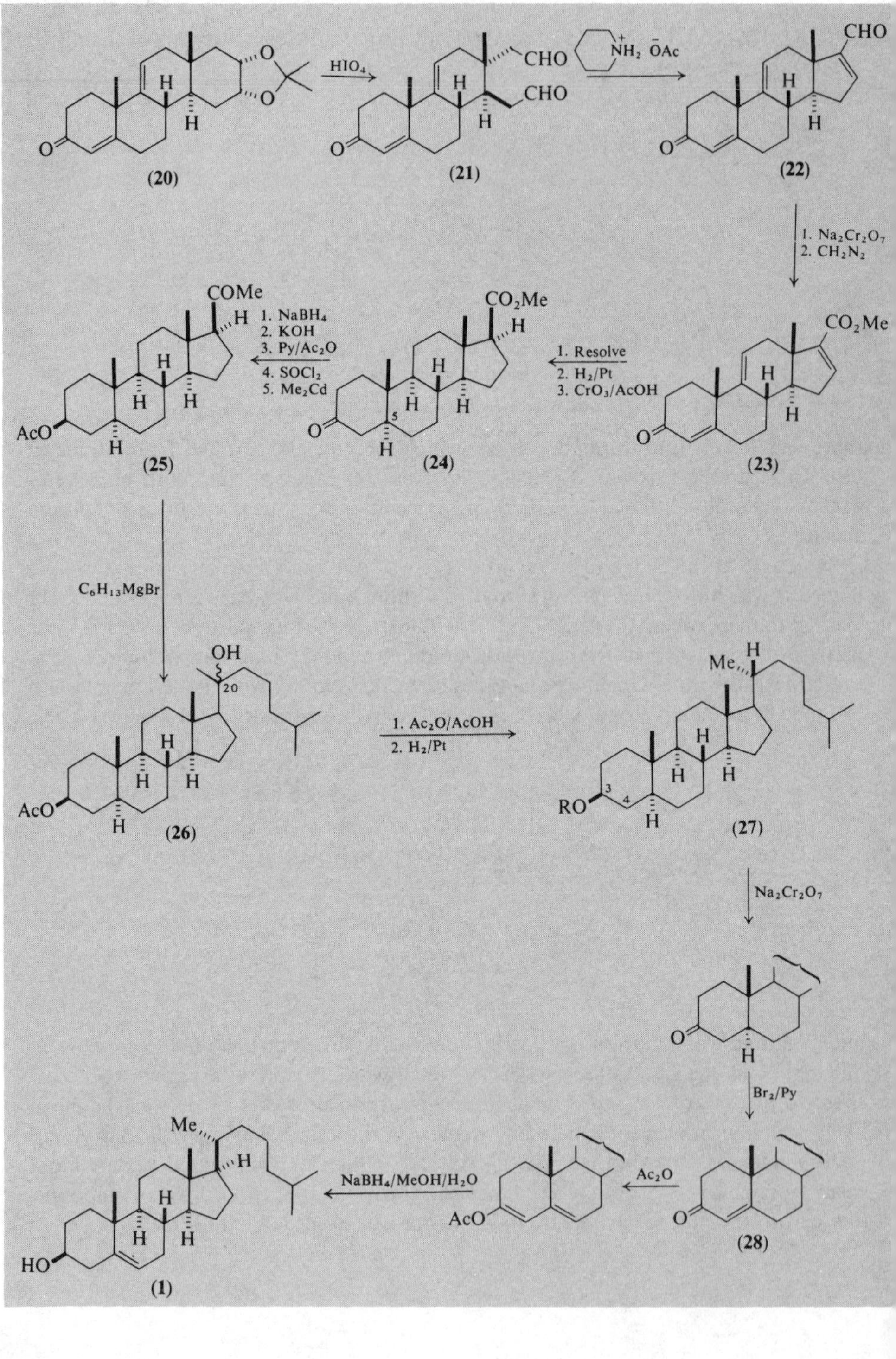

(20)
HIO_4
(21)
$\overset{+}{N}H_2$ $\bar{O}Ac$
(22)
1. $Na_2Cr_2O_7$
2. CH_2N_2
(23)
1. Resolve
2. H_2/Pt
3. $CrO_3/AcOH$
(24)
1. $NaBH_4$
2. KOH
3. Py/Ac_2O
4. $SOCl_2$
5. Me_2Cd
(25)
$C_6H_{13}MgBr$
(26)
1. $Ac_2O/AcOH$
2. H_2/Pt
(27)
$Na_2Cr_2O_7$
Br_2/Py
(28)
Ac_2O
$NaBH_4/MeOH/H_2O$
(1)

Aqueous periodic acid hydrolysed the acetonide of **20** and cleaved the diol to give **21**. Piperidine acetate catalysed the aldol condensation of the dialdehyde to give, as it happens, mainly the desired aldehyde (**22**). This was oxidised to the corresponding acid and then esterified with diazomethane to give **23**, which was resolved (actually by selective precipitation of the 3-β-alcohol using digitonin). The resolved ester (**23**) proved to be identical to material prepared from natural sources and its structure was accordingly beyond doubt. Hydrogenation of **23** took place with delivery of hydrogen to the less hindered α face, so that the configuration at C-9 and C-17 was the natural configuration. (The β face, which carries the two axial methyl groups, C-18 and C-19, is usually the more hindered, but hydrogenation of the C-4 to C-5 double bond is known to be less completely stereoselective and some 5β-H product is probably formed in this case.) Oxidation of the resulting alcohol gave a mixture of ketones (probably the epimers at C-5) from which **24** could be separated.

For the synthesis of cholesterol, the ketone group was reduced (NaBH$_4$), the ester hydrolysed to the acid (KOH), the alcohol acetylated (pyridine and acetic anhydride), and the acid converted to its acid chloride (thionyl chloride) and then treated with dimethyl cadmium to give the ketone (**25**). With excess isohexyl magnesium bromide, **25** gave the two diols (**26**), epimeric at C-20. Acetylation and heat dehydrated the diol at C-20, and hydrogenation then gave a mixture from which cholestanyl acetate (**27**, R = Ac) was isolated. Hydrolysis to the alcohol (**27**, R = H) and conversion of this, by way of cholest-4-en-3-one, to cholesterol itself was already known, and the synthesis of cholesterol was accordingly complete.

For the synthesis of cortisone it was necessary to elaborate the double bond in ring C of **23**. Partial hydrogenation gave the mixture of ketones (**29**) epimeric at C-5. The double bond in ring C was, no doubt, slow to reduce because it was so effectively embedded in the molecule. Reduction of the ketone with sodium borohydride gave a mixture of alcohols from which **30** (R = H) could be isolated. Acetylation gave the acetate (**30**, R = Ac), which was identical to natural material, and this product was hence identified as the desired one. The conversion of the acetate (**30**, R = Ac) to cortisone (**2**) was already known, and the synthesis of cortisone was also complete. The actual sequence involved epoxidation of the double bond and hydrolysis of the acetate and the epoxide groupings to give the 3,9,11-triol. Oxidation then gave the 9-hydroxy-3,11-diketone, isolated as the hemiacetal (**31**). Hydrobromic acid on **31** gave the bromide (**32**, X = Br), which was reduced with zinc to **32** (X = H). Reduction of the 3-keto group, acetylation, conversion of the ester to the acid chloride, and reaction with diazomethane gave the diazoketone (**33**, R = CHN$_2$). Acetic acid gave the acetate (**33**, R = CH$_2$OAc), the cyanohydrin of the monoacetate of which was dehydrated to give **34**. Oxidation of the double bond with osmium tetroxide then gave **35**. Bromination (at C-4) and use of the 2,4-dinitrophenylhydrazone reaction, mentioned in the description of the morphine synthesis, p. 55, gave the 4,5-double bond. Hydrolysis at C-21 completed the synthesis of cortisone.

Questions

1. Make sure that you can write out the mechanism of the complete Robinson ring extension sequence, and, for that matter, the variation used in the construction of ring A.

2. If this was easy to do, convince yourself that the hydrolysis of the protecting group:

was a reasonable process.

3. If this too was easy, look at the conversion of **31** to **32** (X = Br) with HBr. This requires some explanation. Provide one.

References

1. W. E. Bachmann, W. Cole and A. L. Wilds, *J. Amer. Chem. Soc.*, **61**, 974 (1939); **62**, 824 (1940).
2. G. Anner and K. Miescher, *Experientia*, **4**, 25 (1948); *Helv. Chim. Acta*, **31**, 2173 (1948).
3. R. B. Woodward, F. Sondheimer, D. Taub, K. Heusler and W. M. McLamore, *J. Amer. Chem. Soc.*, **74**, 4223 (1952); preliminary communications, *ibid.*, **73**, 2403 3547, 3548, 4057 (1951).
4. H. M. E. Cardwell, J. W. Cornforth, S. R. Duff, H. Holtermann and R. Robinson, *J. Chem. Soc.*, 361 (1953); preliminary communication, *Chem. and Ind.*, 389 (1951).

Cycloartenol

(1)   (2)

Cycloartenol (**1**) is commonly the first intermediate found in plants following the enzyme-catalysed cyclisation of squalene oxide. It is thought to occupy in plants the place in the biosynthetic sequence that lanosterol (**2**) occupies in animals. A synthesis of cycloartenol from lanosterol (**2**) was performed by Barton[1] and would normally be regarded as a *partial synthesis*, a kind of synthesis which is very important and useful but which has not been stressed in this book. In general, partial synthesis is important in relating families of natural products, and especially in determining relative and absolute configurations. It is also useful in that readily available natural products can serve as sources for the commercial manufacture of marketable products, the outstanding example being the conversion of the plant steroids, such as diosgenin, into steroid hormones, such as cortisone.[2] As it happens, since lanosterol (**2**) has been prepared from cholesterol,[3] and the latter has been synthesised many times (see p. 57), the partial synthesis of cyclo-artenol is, formally, a *total* synthesis.

Bromination of lanosterol acetate (to protect the side-chain double bond), and chromic acid oxidation followed by dissolving metal reduction (which also gave back the side-chain double bond) gave the dione (**3**).[4] Wolff–Kishner reduction removed the less hindered 7-keto group, and lithium aluminium hydride reduction gave[5] the diol (**4**), in which the very hindered 11-keto group had been reduced, with hydride attack from the less hindered, lower (α) face of the molecule. Benzo-ylation of the diol gave the mono-3β-benzoate, because the 11β-hydroxyl is very hindered. Nitrosyl chloride in pyridine then gave the 11β-nitrite (**5**). Photolysis of 5 in benzene containing iodine gave the iodoalcohol (**6**), which was immediately oxidised to the iodoketone (**7**).

This reaction, which has acquired the status of a name reaction—*the Barton*

reaction—is the most interesting feature of the synthesis. By its means the function-
ally isolated methyl group has become an iodomethyl group. Such a process, in
which a hydrocarbon residue acquires a heteroatom, is sometimes called a functionalisation reaction. In general, such reactions are observed as random processes in
the chlorination of hydrocarbons and, for example, in the insertion of nitrenes into
C—H bonds. They are occasionally found in the chemistry of bridged polycyclic
terpenoids such as longifolene. Most useful, however, are those cases in which a
specific site on the hydrocarbon grouping can be functionalised, as in the present
synthesis. The mechanism of the reaction is believed to involve light-induced

homolytic cleavage of the nitrite ester, to give the oxy radical (I) and nitric oxide (a 'stable' radical). The radical (I) then abstracts a hydrogen atom from a group

which is spatially well placed, thereby transferring the radical site to carbon (I → II). Radical abstraction reactions of this kind generally show a very high preference for the hydrogen atom six atoms from the radical site and are not dependent on the rigid framework present in this example. The carbon radical (II) abstracts an iodine atom from the iodine, which has been added deliberately. (In the absence of iodine it captures a molecule of nitric oxide and an oxime is isolated.)

The iodoketone (7) with base gave the cyclopropylketone (8), in which the $\Delta^{9(11)}$-enolate has displaced the iodide. Reduction with lithium aluminium hydride *in dioxan* cleaved the ester and completely removed the 11-oxygen function to give cycloartenol (1). The hydrogenolysis is unusual but had been observed in closely related systems when this solvent, with its relatively high boiling point, was used. It seems likely that an intermediate 11β-axial oxygen function was formed and that the loss of this crowded axial group was assisted by the relief of the axial–axial interactions; the fact that the oxygen function is in a cyclopropyl-carbinyl system must also be important.

Question

In the photolysis of **5**, the 19-methyl group becomes functionalised, not the other axial methyl group on ring C, the 18-methyl group. Suggest an explanation, bearing in mind that in Barton's synthesis of aldosterone[6] it is the 18-methyl group which is attacked:

References

1. D. H. R. Barton, D. Kumari, P. Welzel, L. J. Danks and J. F. McGhie, *J. Chem. Soc.* (*C*), 332 (1969).
2. For an account of many of these partial syntheses see L. F. Fieser and M. Fieser, *Steroids*, Reinhold, New York, 1959.
3. R. B. Woodward, A. A. Patchett, D. H. R. Barton, D. A. J. Ives, and R. B. Kelly, *J. Chem. Soc.*, 1131 (1957).
4. W. Voser, O. Jeger and L. Ruzicka, *Helv. Chim. Acta*, **35**, 497 (1952).
5. W. Lawrie, F. S. Spring and H. S. Watson, *Chem. and Ind.*, 1458 (1956).
6. D. H. R. Barton and J. M. Beaton, *J. Amer. Chem. Soc.*, **82**, 2641 (1960).

β-Carotene

(1)

The orange-red pigments, α-, β-, and γ-carotene, so abundantly obvious in carrots, are fat-soluble hydrocarbons possessing ten conjugated double bonds (in α-carotene), eleven conjugated double bonds (in β-carotene) and twelve conjugated double bonds (in γ-carotene). The most symmetrical of these is β-carotene (**1**), which has been synthesised on a commercial scale from β-ionone (**9**), and by many other routes. β-Ionone had itself been synthesised many years before.

One route to β-ionone began with isoprene (**2**), which gave the dibromide (**3**) with hydrobromic acid.[1] When the dibromide was treated with the anion of acetoacetic ester, the primary bromide was displaced and the tertiary bromide lost, by elimination of hydrobromic acid, to give **4**. It is commonly the case that tertiary alkyl halides are not effective in acetoacetic ester reactions, because the rate of the bimolecular elimination reaction with these substrates exceeds the rate of the bimolecular substitution reaction. In the present case, the usual product from a 1,3-dibromide would have been a cyclopropane derivative, which would have been of no use. Alkaline hydrolysis and decarboxylation gave 6-methylhept-5-ene-2-one (**5**). A Reformatsky reaction on this ketone, followed by dehydration of the intermediate *β*-hydroxy ester, gave[2] ethyl geranate (**6**). When the calcium salt of geranic acid was distilled from calcium formate mixed with sand, citral (**7**) was produced. This reaction might be a result of hydride delivery (I) from formate to the carboxylate group of geranate:

(I)

Or it might be a condensation (II) between an enolised geranate and formate, followed by decarboxylation and the return of the double bond to the conjugated

(II)

position. This kind of cookery does not usually give good yields (no yield was quoted in this case) because, although both components do possess features which enable the desired reaction, whichever it is, to take place, neither is particularly well designed for this reaction. Thus the carboxylate anion in I or II is not notably electrophilic; and yet it is required to be so. Nowadays, this reaction would be replaced by a more controlled sequence.

Citral was readily condensed[3] with acetone, in an aldol reaction, to give pseudo-ionone (**8**). On treatment with acid,[4] pseudoionone cyclised to give a mixture of ionones (**9** and **10**), rich in *β*-ionone (**9**). This cyclisation was the result of a carbonium ion attacking a double bond (III), a reaction of a kind we have seen in the

(8) ≡

(III)

(9 and 10)

synthesis of callistephin (p. 26) and will see again, in a more dramatic example, in one of Johnson's steroid syntheses (p. 156). Since citral can be obtained from lemon grass oil, this sequence makes β-ionone readily available.[5]

The remaining steps in the synthesis were worked out by Isler,[6] on the basis of an earlier synthesis by Inhoffen.[7] The β-ionone (9) was converted by the Darzens glycidic ester synthesis to the C_{14} aldehyde (11). In this reaction, the carbon–carbon bond was made by an aldol-like process, but the intermediate (IV) had the opportunity of displacing chloride to give the glycidic ester (V). The ester group was hydrolysed with alkali and the intermediate than decarboxylated (VI), with formation of the aldehyde (11).

(IV)

(11) ← (VI) ← (V)

The aldehyde gave an acetal (12), which was treated with ethyl vinyl ether in the presence of boron trifluoride to give the derivative (13) of a C_{16} aldehyde (14). The reaction with ethyl vinyl ether was a sort of acid-catalysed aldol condensation, the key step of which can be formulated as in VII.

(VII)

The acetal of the C_{16} aldehyde was similarly treated with ethyl propenyl ether to give the derivative (16) of a C_{19} aldehyde (17). A solution of two equivalents of the C_{19} aldehyde was added to bisethynylmagnesium bromide (18), to give the C_{40} diol (19). Acid-catalysed dehydration of this allylic (and propargyllic) dialcohol gave the decaenyne (20), which was reduced, using the Lindlar catalyst so that only the acetylenic bond was attacked. The 15-*cis* isomer (21) of carotene, which was produced in this reduction, was isomerised to the 15-*trans* isomer (1)

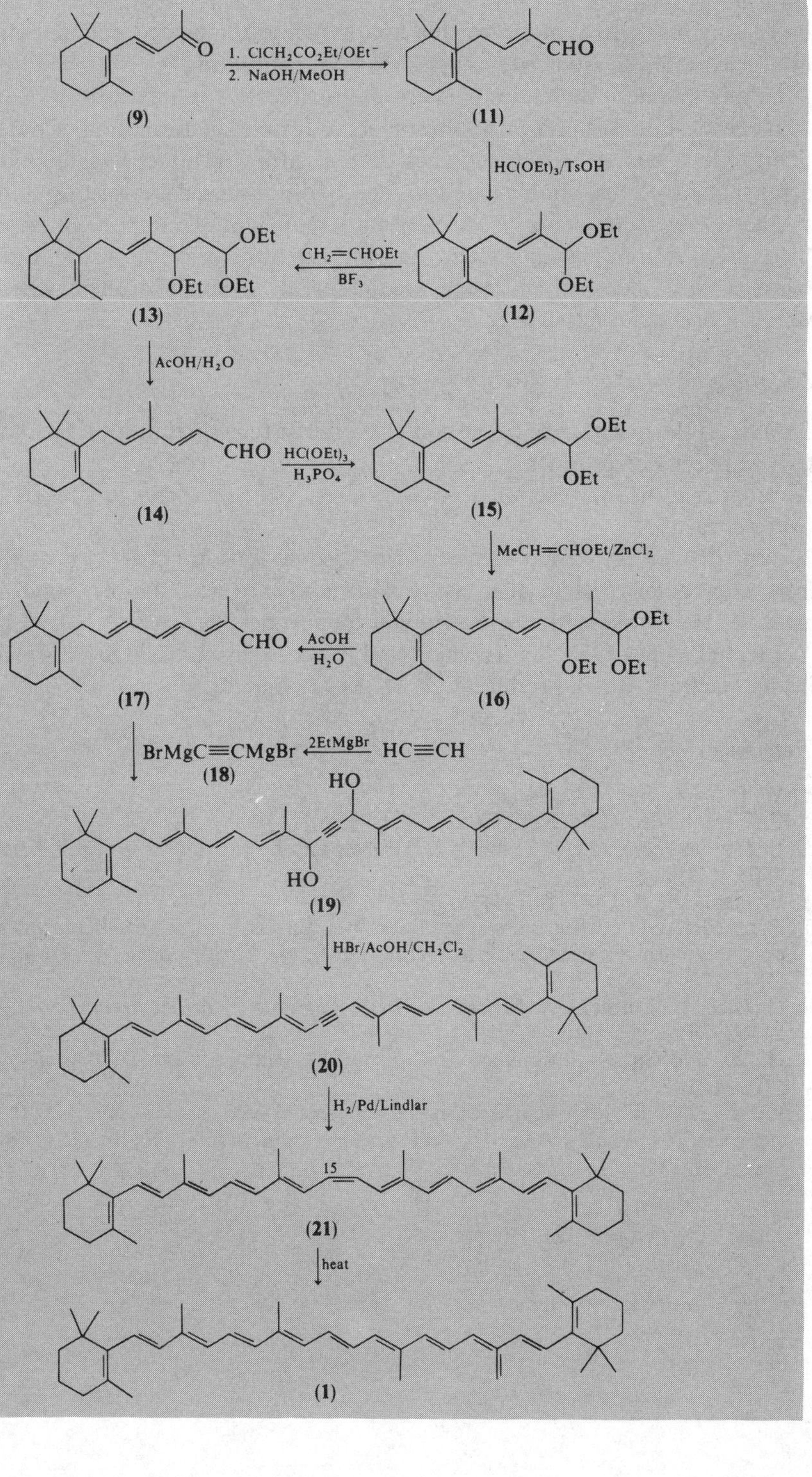
O
1. ClCH2CO2Et/OEt⁻
2. NaOH/MeOH
CHO
(9)
(11)
HC(OEt)3/TsOH
OEt
OEt OEt
CH2=CHOEt
BF3
OEt
OEt
(13)
(12)
AcOH/H2O
CHO
HC(OEt)3
H3PO4
OEt
OEt
(14)
(15)
MeCH=CHOEt/ZnCl2
CHO
AcOH
H2O
OEt
OEt OEt
(17)
(16)
BrMgC≡CMgBr
(18)
2EtMgBr
HC≡CH
HO
HO
(19)
HBr/AcOH/CH2Cl2
(20)
H2/Pd/Lindlar
15
(21)
heat
(1)

by heating it in petrol at 80° for ten hours. A less conjugated double bond would not be expected to isomerise under such mild conditions.

In this synthesis there were many opportunities for mixtures of geometrical isomers and double bond-position isomers to form. The formation of pure crystalline product was at least partly because the natural product had the most stable arrangement of the double bonds. The conditions used in various steps in the synthesis were such as to give these double bonds ample opportunity to get into the most favourable arrangement; and the reactions proved, on the whole, to be relatively well behaved in giving predominantly a single product when several were, in principle, possible.[8]

Questions

1. What is the mechanism of the addition of HBr to isoprene so that it should lead to **3** as the major product?

2. What reagent, or reaction sequence, might be used today to convert the ester (**6**) to citral (**7**)?

3. Why do you think that the *cis* → *trans* isomerisation (**21** → **1**) is easier in this long conjugated system than it is with unconjugated double bonds? Iodine catalyses the isomerisation: how do you think it acts?

4. Identify some of the factors which lead to the isolation of largely a single isomer in the reactions which produce **8**, **9**, **11**, **14**, **17** and **20**.

References

1. W. Ipatiew, *Ber.*, **34**, 594 (1901).
2. F. Tiemann, *Ber.*, **31**, 825 (1898).
3. F. Tiemann, *Ber.*, **32**, 115 (1899); H. Hibbert and L. T. Cannon, *J. Amer. Chem. Soc.*, **46**, 119 (1924).
4. F. Tiemann, *Ber.*, **31**, 870 (1898).
5. For a review of cyclisation of pseudoionones, discussing the proportions of α- and β-products with different acidic reagents, see E. E. Royals, *Ind. Eng. Chem.*, **38**, 546 (1946).
6. O. Isler, H. Lindlar, M. Montavon, R. Ruegg and P. Zeller, *Helv. Chim. Acta*, **39**, 249 (1956).
7. H. H. Inhoffen, F. Bohlmann, K. Bartram, G. Rummert and H. Pommer, *Annalen*, **570**, 54 (1950).
8. For a review of other syntheses in the carotenoid field, see, O. Isler and P. Schudel, *Advances in Organic Chemistry, Methods and Results*, Interscience, New York, 1963, Vol. 4, p. 115.

Dehydroabietic Acid

(1)　　　　　　　　　　　(2)

Dehydroabietic acid (1) is a constituent of the oleoresin of *Pinus palustris*, but can more readily be got from abietic acid (2), the diterpene most easily isolated from rosin. After the completion of the synthesis[1] of 1 in 1956, its conversion to 2 was effected[2] by lithium-in-ethylamine reduction, followed by isomerisation of the double bonds under acidic conditions. Abietic acid itself (2) is not actually an abundant compound in fresh oleoresins, but the mixture of the several dienes which are present is largely transformed into abietic acid under acidic conditions.

The synthesis[1] began with 2-acetylnaphthalene (3), which was converted to 2-isopropylnaphthalene (4). Sulphonation followed by potash fusion gave 6-isopropyl-2-naphthol, which was reduced by the method of Birch to give the tetra-lone (5).

The next stage, the formation and alkylation of the enamine (6), is a special story. The need for monomethylation at this stage had led Stork to search for a new method. He found that the reaction of methyl iodide with 2-tetralone enamines such as 6 did give, after hydrolysis, high yields of the monomethyl derivative (7); the more conventional use of alkoxide and methyl iodide on the ketone itself gave largely the dimethyltetralone. That enamines are useful reagents for mono-alkylation is now well known;[3] their reactions, since Stork's original work,[4] have been much exploited. Curiously, 2-tetraolones are most unusual among ketones in giving smooth monomethylation by this method; in general, alkylation of enamines with rather weakly electrophilic alkyl halides, is not preparatively useful. Monoalkylation of enamines by electrophilic olefins, such as methyl acrylate or acrylonitrile, is much more general (see p. 178).

Five more carbon atoms were now added to the ketone (7), using a Robinson

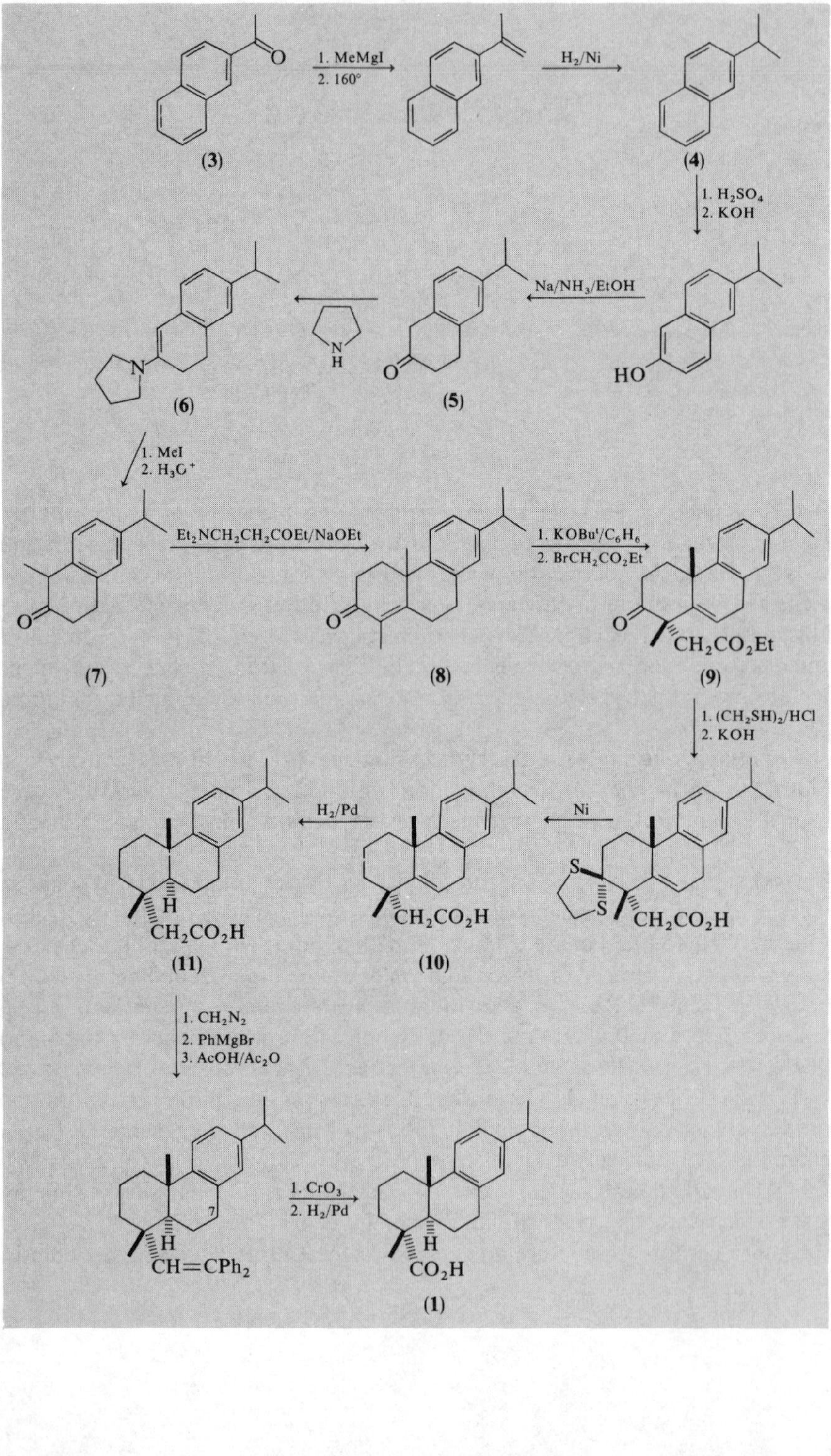

O
1. MeMgI
2. 160°
H_2/Ni
(3)
(4)
1. H_2SO_4
2. KOH
Na/NH_3/EtOH
HO
N
H
O
N
(6)
(5)
1. MeI
2. H_3O^+
$Et_2NCH_2CH_2COEt$/NaOEt
1. $KOBu^t/C_6H_6$
2. $BrCH_2CO_2Et$
O
O
O
CH_2CO_2Et
(7)
(8)
(9)
1. $(CH_2SH)_2$/HCl
2. KOH
H_2/Pd
Ni
S
S
H
CH_2CO_2H
CH_2CO_2H
CH_2CO_2H
(11)
(10)
1. CH_2N_2
2. PhMgBr
3. AcOH/Ac_2O
7
1. CrO_3
2. H_2/Pd
H
H
$CH{=}CPh_2$
CO_2H
(1)

ring extension (see p. 59) with ethyl vinyl ketone generated *in situ* from the Mannich base. In spite of the presence in the product (**8**) of 19 of the 20 carbon atoms present in dehydroabietic acid, stereochemical problems had so far been avoided. The next step introduced the second chiral centre: the ketone (**8**), with potassium t-butoxide, gave an enolate which was combined with ethyl bromoacetate. The product was **9**.

That the alkylation of the dienolate system (I and II) would take place at C-4 rather than C-6 was well known from many analogies in steroid and terpenoid chemistry (see p. 61 for one example). That the newly introduced group would approach from the α-surface, was less certain. In general, alkylation of such an enolate system (I and II) involves attack at right angles to the double bond in such

a way as to develop head-on overlap with the p-orbital on C-4. Attack from the β-direction (I) obviously has the disadvantage that the bromoester interacts, in a 1,3-diaxial fashion, with the axial methyl group; α-attack (II) avoids this interaction. However, as α-attack proceeds, ring A must buckle in such a way that C-3 moves up and C-4 down—that is towards a boat conformation. β-Attack does not have this undesirable effect, because the buckling in this case is towards a chair conformation. The stereochemical outcome of such a reaction is therefore a delicate balance between these unfavourable factors. In this particular case, with a relatively bulky alkyl group being introduced, the former factor was evidently the more important and, as was expected on the basis of somewhat similar reactions in the literature, α-attack occurred.

The ketone group had now served its purpose and was removed by forming the thioketal and desulphurising with Raney nickel. The double bond was hydrogenated and once again a cleanly stereoselective delivery of hydrogen to the α-surface was expected and observed. The rigid system of **10** forced the two methyl groups in the vicinity of the double bond to be axial; the α-acetic acid side-chain, although larger than a methyl group, was equatorially disposed and was therefore much less effective in hindering approach (or adsorption) in that direction.

The side-chain had, at this stage, one carbon too many. The removal of one of these carbons was readily achieved using the Barbier-Wieland degradation, a reaction which, before the universal adoption of physical methods, had found much use in the determination of the structure of natural products. The methyl ester of **11**, with phenyl magnesium bromide followed by dehydration, gave the olefin (**12**). Oxidation with chromic acid gave simultaneous oxidation at the double bond and at the benzylic position C-7, but the crude product was deprived of this extra functionality by hydrogenation and hydrogenolysis, which gave racemic dehydro-abietic acid (**1**).

This work is a very pleasing and comparatively early example, illustrating two of the more enjoyable aspects of synthesis: development of a new method for a specific purpose, and the application of both analogy and mechanistic insight in choosing stereoselective reactions. The elegance of the method is shown by the production of a compound possessing three chiral centres without the necessity of separating and identifying a pair of diastereoisomers at any stage.

Question

1. Enamine alkylations go because of the overlap of the nitrogen lone pair with the double bond, just as the negative charge of an enolate makes the β-carbon nucleophilic:

(V)

Consider, therefore, why enamines are more easily persuaded to stop at mono-alkylation. It might be simpler to consider cyclohexanone and 2-alkylcyclohexanone, rather than **5** and **7**. It may also be helpful to know that the stable enamines prepared from 2-alkyl cyclohexanones are largely, though not exclusively, the tautomers of type V.

References

1. G. Stork and J. W. Schulenberg, *J. Amer. Chem. Soc.*, **84**, 284 (1962).
2. A. W. Burgstahler and L. R. Worden, *ibid.*, **83**, 2587 (1961).
3. A. G. Cook (Ed.), *Enamines*, Marcel Dekker, New York, 1969.
4. G. Stork, A. Brizzolara, H. Landesman, J. Szmuszkovicz and R. Terrell, *J. Amer. Chem. Soc.*, **85**, 207 (1963).

The Penicillins

RHN $\overset{H}{\underset{}{\rule{0pt}{0pt}}}\overset{H}{\underset{}{\rule{0pt}{0pt}}}$ S

(1)

structure with CO_2H, N, O positions

R = PhOCH$_2$CO	Penicillin V
R = PhCH$_2$CO	Penicillin G
R = ⌒⌒⌒CO	Penicillin F
R = HO-p-C$_6$H$_4$CO	Penicillin X
R = ⌒⌒⌒⌒CO	Penicillin K

The chemistry of penicillin was investigated in England and the United States in the last years of the second world war in an unparalleled effort to derive a structure. In the structural work an armoury of physical methods, unprecedented at the time, was brought to bear on the problem. X-ray crystallography, infrared spectroscopy, the measurement of dissociation constants, and thermochemical data were particularly important in confirming the structure (1), which was one of three possibilities suggested by the degradative work.[1] One hope which spurred on the investigation was that the penicillins might prove to have a structure simple enough for synthesis to be commercially feasible. As it turned out, although the structure is fairly simple, a synthesis was by no means easy to find. Penicillin was obtained[2] this way in only trace amounts before the first effective synthesis was achieved[3] by Sheehan in 1957. In the intervening period, microbiological technology made the production of penicillins from natural sources so cheap that synthesis was never able to compete.

The difficulty was the problem of closing a β-amino acid to a β-lactam. This particular β-lactam is unusually reactive: the bicyclic system constrains the molecule and reduces the overlap of the nitrogen lone pair with the carbonyl group; the carbonyl group is more subject to nucleophilic attack than in normal amides, or normal β-lactams, and the ring is, consequently, readily opened, particularly in acidic conditions.[4] The solution of the problem came with the development[5] of carbodiimides as cyclising reagents. Another feature of the synthesis was the straightforward use of protecting groups (phthalimido for the amine and t-butyl for one of the carboxyls) which were smoothly removed when required.

Valine (2) was chloroacetylated and the intermediate (3) cyclised to the oxazolone (4). The double bond in this product did not appear in the most obvious of places. Cyclisation to the intermediate (I) was normal and the ready formation of the enolate (II) might have been expected to lead to the isomer of **4** namely III

Evidently the rearrangement of the double bond to give the more substituted isomer (4) was easy under the reaction conditions. Such rearrangement could be acid or base catalysed and was perhaps not too surprising.

A similar rearrangement takes place in the course of the Sommelet reaction: the ylid (IV), for example, gives the non-aromatic isomer (V), which rearranges, with restoration of the benzene ring, in what is probably, in this case, a base-catalysed reaction.

(IV)   (V)

With hydrogen sulphide and methoxide the oxazolone (**4**) gave the acetyl derivative (**5**), which gave racemic penicillamine (**6**) on hydrolysis.

Glycine was converted to the protected derivative (**7**), which gave, with t-butoxide and t-butyl formate, the Claisen condensation product (**8**). Combination of **6** with **8** gave the protected derivative (**9**) of penicilloic acid together with a diastereoisomer (**10**). In this reaction two new chiral centres are set up, but only two diastereoisomers were obtained, known as the α- and γ-isomers. Evidently the two substituents (the carboxyl group and the glycine residue) on the thiazolidine ring were *trans* to each other in both of these isomers because the isomers were equilibrated in pyridine. Such equilibration could only occur at the carbon next to the carboxylate ester group, which can readily enolise.

The two isomers were separated by fractional crystallisation, and since each could be equilibrated, it was possible to convert the unwanted isomer, at least in part, into the desired one and thus prevent waste. The subsequent synthesis showed that the higher-melting isomer, which they had called the α-isomer, was the one they wanted. The phthalimide group was removed from the α-isomer (**9**) with hydrazine and the amine hydrochloride was re-acylated to give a side-chain which is one of those found in the natural penicillins: the phenoxyacetyl group. The t-butyl group was removed with dry hydrogen chloride in methylene dichloride, and the amino acid (**11**) was liberated from the hydrochloride by neutralisation with pyridine. Cyclisation was carried out on the mono sodium salt with dicyclo-hexyl carbodiimide in aqueous dioxan; the penicillin V (**12**) was extracted into ether and then isolated as its potassium salt. It was resolved with (−)-*erythro*-1,2-diphenyl-2-methylaminoethanol, which happened to be available because the racemic amine could be resolved with penicillin G! The yield in the cyclisation step was only 5 per cent.

In subsequent work,[6] other protecting groups have been used at various stages, with consequent improvement in both yield and generality. For example, benzyl ester formation from (**9**) followed by the removal of the other two protecting groups gave the amino acid (**13**). Protection of the amino group by tritylation gave **14**, which could be cyclised with diisopropyl carbodiimide to give **15** in 67 per cent yield.

In the earlier work, when the benzyloxycarbonyl group was used, oxazolone (azlactone) formation (VI) was a side reaction in the cyclisation step; this common pitfall in peptide synthesis (see p. 99) was avoided when the trityl group was used, and consequently the yield of β-lactam was much improved. Moreover, the benzyl

and the trityl group can be removed, by hydrogenolysis and careful treatment with aqueous acid respectively, to give 6-aminopenicillanic acid (**16**, R = H). This amine can be acylated with a wide variety of groups, some of which give the natural penicillins, for example V (**12**) or G (**16**, R = $PhCH_2CO$), while others give useful 'semi-synthetic' penicillins.[7]

Questions

1. What is the mechanism of the cyclisation of **3** to give the presumed intermediate (I)? What do you think the most likely mechanism for the isomerisation of the oxazolone (III) to the oxazolone (**4**)?

2. The carboxylate anion of the penicilloic acid (11) attacks the carbodiimide at the central carbon atom to give an intermediate (VII). Identify the features of this intermediate which lead it to undergo cyclisation to the β-lactam and to the oxazolone (VI).

(VI)   (VII)

References

1. The account of all the war-time work leading to the structure of the penicillins was published in book form: H. T. Clarke, J. R. Johnson and R. Robinson (Eds.), *The Chemistry of Penicillin*, Princeton University Press, Princeton, N.J., 1949. For a review of this work see also A. H. Cook, *Quart. Rev.*, **2**, 203 (1948).
2. See K. Folkers in A. R. Todd (Ed.), *Perspectives in Organic Chemistry*, Interscience, New York, 1956.
3. J. C. Sheehan and K. R. Henery-Logan, *J. Amer. Chem. Soc.*, **81**, 3089 (1959).
4. R. B. Woodward, in reference 1, pp. 443–449.
5. J. C. Sheehan and G. P. Hess, *J. Amer. Chem. Soc.*, **77**, 1067 (1955).
6. J. C. Sheehan and K. R. Henery-Logan, *J. Amer. Chem. Soc.*, **84**, 2983 (1962).
7. A short book is devoted to an account of these and other syntheses of penicillins and their derivatives M. S. Manhas and A. K. Bose, *Synthesis of Penicillin, Cephalosporin C and Analogs*, Marcel Dekker, New York, 1969.

Cephalosporin C

(1)*

Like the penicillins, which they resemble closely in structure, the cephalosporins are antibiotics with a wide range of activity. Cephalosporin C (1) is itself relatively inactive, although several related compounds, with other side chains, have high antimicrobial activity. It is particularly important because the cephalosporins remain active against organisms which have become resistant to the penicillins. Although the cephalosporins are not yet much used in therapy, they could become a second line of defence.

The structural resemblance of cephalosporins to penicillins made a synthesis along the same line as Sheehan's synthesis of penicillins a tempting route. Woodward chose not to follow it.[1] The knowledge that the product from the cleavage of the β-lactam ring was not available, as it had been in the penicillin series, was a decisive factor. All attempts to hydrolyse the β-lactam of cephalosporins had caused further degradation, and it was obviously unwise to aim for the synthesis of an intermediate which would probably be even less stable than the cephalosporins themselves. Instead, Woodward chose to begin by constructing the β-lactam ring, the very feature which had been left until last in the synthesis of the penicillins. This strategy had the advantage that what was potentially the hardest part of the synthesis would be done early; the six-membered ring, which would be made later in the synthesis, would probably not present any special problems. It had the inevitable disadvantage that the most sensitive feature of the molecule would have to be carried through several stages of the synthesis—the second ring would have to be prepared, equilibrations carried out, and protecting

---

* The drawing on the left is for comparison of the structure of cephalosporin C with that of penicillin on p. 82; this is the usual way of drawing the penicillins. The drawing on the right is the way Woodward draws cephalosporin C. To make easy comparison of these pages with the original, I have used the latter orientation throughout.

groups put on and taken off, all with the β-lactam ring present in the molecule. A third factor was borne in mind in choosing this strategy. Because the β-lactam ring is common both to the penicillins and to the cephalosporins, there is strong evidence that its presence is essential for their antimicrobial activity; by starting with the β-lactam ring, Woodward would have the opportunity to build other rings onto it which would be similar but not identical to those present in the penicillins and cephalosporins. Hitherto, this kind of structural variation had not been easy— structural variation had been confined largely to changes in the side chains. There was, therefore, the hope that unnatural, but more potent, therapeutic agents could be prepared.

The starting material was L(+)-cysteine (2), which was protected at three of its reactive sites as the derivative (3). We shall see that the key to success lay in this careful use of protecting groups. The next step had little precedent. It was necessary to introduce a nitrogen atom onto C-6 in 3 (cephalosporin numbering); but that carbon atom is not, as it stands, particularly nucleophilic or electrophilic. However, there is a useful looking but little known, reaction in the literature:[2] tertiary amines (I), in which one group is a methyl group, react with diethyl azodicarboxylate to give an adduct (IV), which is readily hydrolysed to give a secondary amine (V), formaldehyde (VI), and hydrazodicarboxylate (VII). In this reaction, the

transformation (I → IV) is of the kind needed for the functionalisation of the methylene group of 3; the only difference is that 3 has a sulphur atom as the potential site of initial reaction, whereas I has a nitrogen atom.

When 3 was heated with dimethyl azodicarboxylate, the product was indeed the hydrazodiester (4), analogous to the hydrazodiester (IV). (The stereoselectivity observed was presumably a result of attack on the less hindered side of the thia-zolidine ring.) The hydrazodiester (**4**) was oxidised with lead tetraacetate, and the mixture was then treated with sodium acetate in boiling methanol. These reagents probably caused, first, the oxidation of the hydrazo group to an azo group (VIII), second, ready loss (probably by attack of acetic acid) of the methoxy-carbonyl group (VIII, arrows), hydrogen abstraction (IX) (from the site which would generate the most stable radical), combination with an acetoxy radical (X), acetolysis (or acid-catalysed methanolysis) of the azoester (XI), and finally, acetate-catalysed methanolysis to give the alcohol (XV = **5**). Once again, the newly

introduced hydroxyl group was found to be *trans* to the methoxycarbonyl group; there was no trace of the *cis* isomer (XVII). Yet, if the reaction sequence is stopped at the ester (XIII and XIV) stage, a small amount of the *cis* ester (XIII) is present. Apparently, any *cis* alcohol (XVII) produced from the *cis* ester is converted to the more stable *trans* alcohol (XV). This equilibration (XV ⇌ XVI ⇌ XVII) is most likely to go by way of an opening of the thiazolidine ring:

It is remarkable that no drastic decomposition should accompany the opening and closing of the ring. The capacities for self destruction of the open chain tautomer (XVI) (and also of the alcohols themselves) are many and obvious. But thioacetone was not produced, and no racemisation occurred in spite of the known ease with which α-formyl esters (such as XVI) enolise. Evidently the opening and closing reaction was the only one going on.

The hydroxyl group in **5** was on the opposite side of the ring from that desired for the nitrogen atom. Mesylation and displacement of the mesylate by azide ion introduced the nitrogen function with inversion of configuration. There was some risk that the displacement would be of the unimolecular (SN1) type (XVIII → XX)

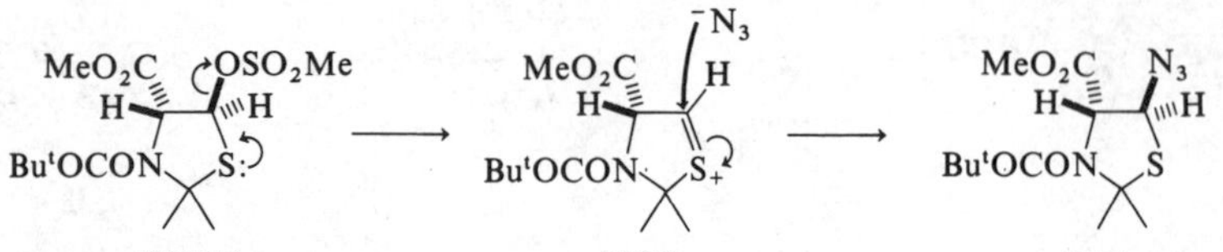

$$\text{(XVIII)} \qquad \text{(XIX)} \qquad \text{(XX)}$$

with assistance by sulphur, and might therefore give, stereoselectively, the *trans* product. This did not occur because sulphur (cf. XXI) is not as good as oxygen (cf. XXII) at supporting a positive charge on an adjacent carbon atom.[3]

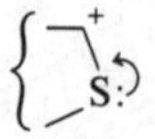

$$\text{(XXI)} \qquad\qquad \text{(XXII)}$$
$$\text{poor overlap} \qquad \text{good overlap}$$

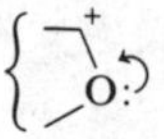

The azide (**6**) was reduced to the amine (**7**) with aluminium amalgam. The more usual hydrogenation could not be used because sulphur was present in the molecule. The β-lactam ring was then closed, using triisobutyl aluminium, which had not been used for this purpose before. The mechanism of this reaction is not known, but we may see in the reactivity of an organometallic compound like triisobutyl aluminium the capacities needed for such a reaction. The aluminium atom can function as a Lewis acid, and the metal-alkyl bonds can function as basic sites:

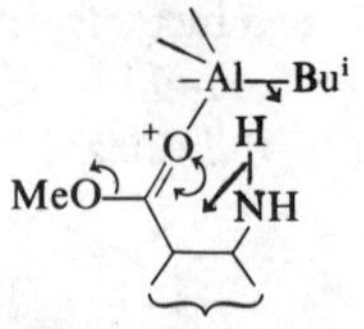

The β-lactam (**8**) was then complete, and the stage was set for joining it to the carbon atoms which would make up the six-membered ring. The compound used to provide these three carbon atoms was the dialdehyde (**14**), which was prepared

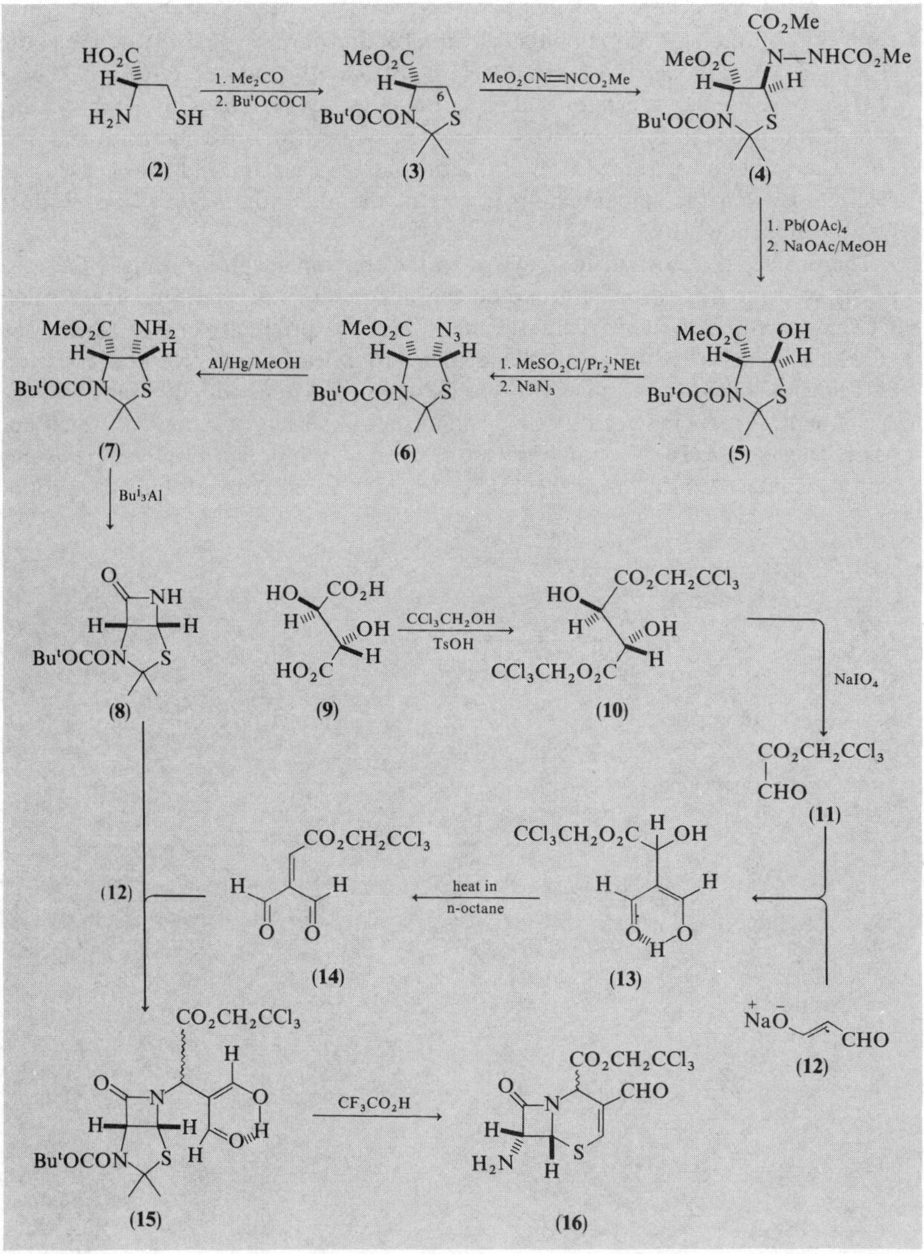
(2)
1. Me_2CO
2. Bu^tOCOCl
(3)
$MeO_2CN{=}NCO_2Me$
(4)
1. $Pb(OAc)_4$
2. $NaOAc/MeOH$
(5)
1. $MeSO_2Cl/Pr_2{}^iNEt$
2. NaN_3
(6)
$Al/Hg/MeOH$
(7)
Bu^i_3Al
(8)
(9)
CCl_3CH_2OH
$TsOH$
(10)
$NaIO_4$
(11)
(12)
heat in
n-octane
(13)
(14)
(12)
(15)
CF_3CO_2H
(16)

from the sodium salt of malondialdehyde (12) and tartaric acid (9) by the route shown. The combination of the β-lactam (8) and the very reactive dialdehyde (14) gave, without any need for acid or base catalysis, the adduct (15). At this point, two of the protecting groups were dispensed with: trifluoroacetic acid removed both the t-butyloxycarbonyl group and the acetonyl group, and also caused the cyclisation from the sulphur atom to one of the aldehyde groups. The product was the aldehyde (16).

The protected side chain was joined to the free amino group using DCC (see p. 84), and the free carboxyl group on the side chain was esterified, again using DCC, with β,β,β-trichloroethanol. There were two products, one of which was the aldehyde (17). Reduction of the aldehyde group with diborane and acetylation of the product gave the acetate (18). The reduction was done with diborane because it is a neutral reducing agent, unlike sodium borohydride, solutions of which are basic. The acetate (18) was equilibrated, in pyridine solution at room temperature, with its isomer (19), in which the double bond is conjugated with the ester group.

This isomer was separated out and the other isomer re-equilibrated. The conjugated isomer (19) was then freed from the β,β,β-trichloroethyloxy groups by reduction with zinc in aqueous acetic acid at 0°. This protecting group was a new one. It was removed easily, partly because the presence of three chlorine atoms provides a favourable statistical factor, and partly because the reductive removal of one of them (XXIII) is assisted by the electron withdrawing effect of the other two.

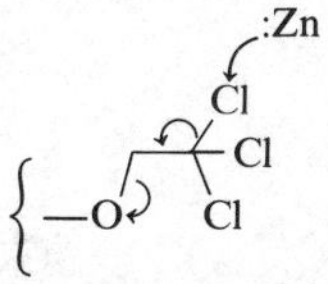

(XXIII)

Following its appearance in this synthesis, the β,β,β-trichloroethyloxy protecting group has been used a number of times. Its special value, demonstrated in this synthesis, is the ease with which it can be removed under conditions which are mild enough not to affect other sensitive sites in the molecule.

Questions

1. The rearrangement of III to IV is of a kind well precedented in the Wittig and the Stevens rearrangements. What is the mechanism?
2. How would you synthesise the α-aminoadipic acid which makes up the side chain?

References

1. R. B. Woodward, K. Heusler, J. Gosteli, P. Naegeli, W. Oppolzer, R. Ramage, S. Ranganathan and H. Vorbrüggen, *J. Amer. Chem. Soc.*, **88**, 852 (1966); R. B. Woodward, *Science*, **153**, 487 (1966).
2. O. Diels and E. Fischer, *Ber.*, **47**, 2043 (1914).
3. For a discussion of this point see R. L. Autrey and P. W. Scullard, *J. Amer. Chem. Soc.*, **90**, 4924 (1968).

Coenzyme A

(1)

The carriers of acyl groups in biochemical systems are the *S*-acyl derivatives of coenzyme A. The structure of coenzyme A **(1)** presents one of the most complex synthetic problems in nucleotide chemistry. The structural work was complete in the early 1950's; it is remarkable that at that time even the best preparations were still only 60% pure. The first synthesis[1] was reported in 1959. This synthesis, like those we have just been examining, involves the deployment of a number of protecting groups and also the use of special reagents for condensation.

Pantetheine **(8)** had already been synthesised by the following route.[2,3] iso-Butyraldehyde and formaldehyde with base gave the mixed aldol product **(2)**, which, via the cyanohydrin and hydrolysis, gave pantolactone **(3)**. This was resolved by alkaline hydrolysis, formation of an insoluble quinine salt, and regeneration of the lactone by acidification of the salt. The more soluble quinine salt gave the unwanted (+)-lactone, but this could be recycled by racemising it with hot alkali. The (−) lactone was reacted with the amine **(7)** to give pantetheine **(8)**. The amine **(7)** had itself been prepared by a conventional peptide synthesis involving: (i) *protection* of the sulphydryl group **(4)**; (ii) *protection* of the amino group of *β*-alanine **(5)**; (iii) *activation* of the carboxyl group of the *β*-alanine, in this case as the azide **(5)**, which was obtained from the hydrazide with nitrous acid; (iv) *condensation* to give **6** as a result of bond formation between the only nucleophilic group remaining and the activated acid; and (v) *removal of the protecting groups* to give the thiol **(7)**.

Pantetheine **(8)** is unstable in the air and gave the disulphide **(9)**, known as pantethine. The alcohol **(9)** was phosphorylated in dry pyridine with dibenzyl-phosphorochloridate **(10)** (made from dibenzylphosphite and *N*-chlorosuccinimide).

In the presence of a suitably placed hydroxyl group, phosphate triesters are uncomfortably labile, being very easily converted into cyclic phosphates. Direct reductive removal of the protecting groups from **11** did indeed result in substantial quantities of the cyclic phosphate (**14**), in addition to the required monophosphate (**13**). However, mild hydrolysis to the dialkyl phosphate (**12**), followed by reductive removal of the second benzyl group, gave **13** in good yield.

The nucleotide part of coenzyme A was prepared from adenosine (**20**), which has been prepared in a number of ways. A typical route would be to protect the amino group of adenine (**15**), make the mercury derivative (**16**), and combine this fragment, nucleophilic at N-9, with the protected bromosugar (**18**), which is prepared from tetraacetylribose (**17**) with hydrobromic acid. The stereochemistry at the glycosidic position, C-1′, is assured because bromosugars, with an acetyl group *trans* on the C-2′ hydroxyl react in the usual solvent, boiling xylene, with neighbouring group participation (I → III). Acetobromosugars (as compounds of type **18** are usually called) epimeric at C-1′ react with simple inversion; so the natural *trans* arrangement of substituents on C-1′ and C-2′ is the most simple to obtain. Removal of the acetyl groups gives adenosine.

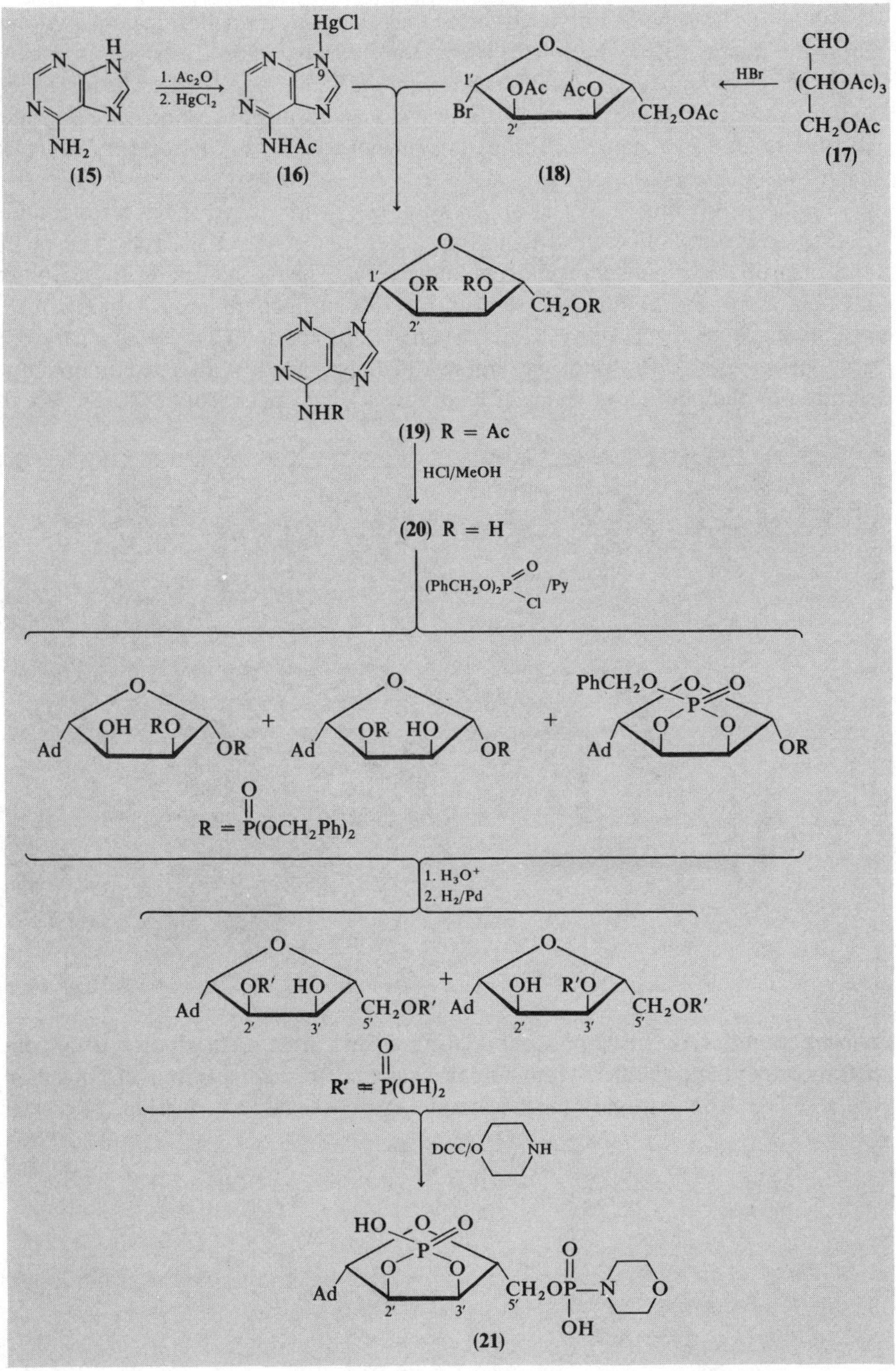
1. Ac2O
2. HgCl2
HBr
HCl/MeOH
(PhCH2O)2P(O)Cl /Py
R = P(O)(OCH2Ph)2
1. H3O+
2. H2/Pd
R' = P(O)(OH)2
DCC/morpholine
(15)
(16)
(17)
(18)
(19) R = Ac
(20) R = H
(21)

Adenosine was phosphorylated with dibenzylphosphorochloridate in pyridine to give a mixture of triphosphates, which on hydrolysis and hydrogenolysis gave the expected mixture of 2′,5′-diphosphate and the 3′,5′-diphosphate. The formation and hydrolysis of cyclic phosphates derived from a *cis*-diol are especially easy when the diol is on a five membered ring; this phenomenon is an important feature of nucleotide chemistry.[4]

Treatment of this mixture of diphosphates with dicyclohexylcarbodiimide (DCC) and morpholine caused the formation first of the 2′,3′-cyclic phosphate and then of the 5′-phosphoromorpholidate (21). (The mechanism of the action of DCC in activating a phosphoryl group towards nucleophilic attack by hydroxyl or amino groups is, of course, closely similar to its action in activating a carboxyl group, see p. 84.) The phosphoramidate (21) was readily attacked in pyridine solution by the phosphate anion of 13 to give the pyrophosphate (22). The use of

coenzyme A (1) + iso-coenzyme A

phosphoramides as active phosphorylating agents must seem strange to anyone extrapolating from the behaviour of carboxamides to the behaviour of phosphoramides. The former are not good acylating agents because the overlap (IV) serves to reduce both the electrophilicity of the carbonyl group and the basicity of the amino group. With phosphoramides, however, the corresponding overlap is much less important; the amino group is readily protonated and becomes a good leaving group (V).

The cyclic phosphate in **22** was opened with dilute acid to give, as usual, the mixture of 2'- and 3'-phosphates, which were then separated by ion exchange chromatography. The 3'-phosphate was coenzyme A.

References

1. J. G. Moffatt and H. G. Khorana, *J. Amer. Chem. Soc.*, **83**, 663 (1961); summarised in A. M. Michelson, *The Chemistry of Nucleosides and Nucleotides*, Academic Press, New York, 1963.
2. E. T. Stiller, S. A. Harris, J. Finkelstein, J. C. Keresztesy and K. Folkers, *J. Amer. Chem. Soc.*, **62**, 1785 (1940).
3. J. Baddiley and E. M. Thain, *J. Chem. Soc.*, 800 (1952); E. E. Snell and his co-workers, *J. Amer., Chem. Soc.*, **75**, 1694 (1953).
4. A. J. Kirby and S. G. Warren, *The Organic Chemistry of Phosphorus*, Elsevier, Amsterdam, London and New York, 1967, chapter 10.
5. For a discussion of this point, see D. M. Brown in *Advances in Organic Chemistry, Methods and Results*, Interscience, New York, 1963, Vol. 3, p. 75.

Peptide Synthesis: Bradykinin

ARG–PRO–PRO–GLY–PHE–SER–PRO–PHE–ARG

(1)

The synthesis of peptides[1] presents very special problems. The usual procedure is to start with the individual amino acids and to concentrate on the making of the-amide bonds. The synthesis of a nonapeptide—like bradykinin, with eight equilibrateable chiral centres—does not present the problem we have had up to now—of generating chiral structures stereoselectively,—but presents rather the problem of maintaining the configuration of chiral centres through a large number of reactions. Much knowledge about how to do this has been built up in the course of synthetic work on peptides, and many reagents have been developed to perform one or more stages of a sequence resulting in the joining of a carboxylic acid group of one amino acid to the amino group of another:

$$\text{(I)} \qquad + \qquad \text{(II)} \qquad \longrightarrow$$

Of course this can be done just by heating: amino groups are nucleophilic and carboxyl groups are electrophilic. But these conditions would cause extensive racemisation when applied to amino acids, and no control would be exercised over the number of peptide bonds which are formed, or over which amino acid combined with which: I with I, II with II, or I with II.

Thus the usual procedure is (i) to protect both the amino group and the carboxylic acid group not involved in the desired reaction, and (ii) to convert the carboxylic acid group to a more electrophilic group. The combination of the protected derivatives:

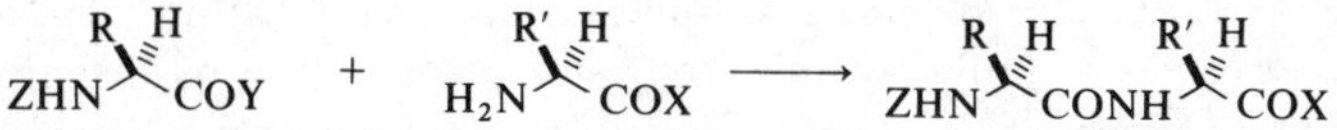

$$ZHN \overset{R}{\underset{}{\overset{H}{\diagup}}} COY \quad + \quad H_2N \overset{R'}{\underset{}{\overset{H}{\diagup}}} COX \quad \longrightarrow \quad ZHN \overset{R}{\underset{}{\overset{H}{\diagup}}} CONH \overset{R'}{\underset{}{\overset{H}{\diagup}}} COX$$
$$(III)$$

is then controllable and selective, and the removal of the protecting groups gives the desired product. It is in the choice of the groups **X**, **Y** and **Z** that skill and experience tell. It is particularly useful if the group **Z** in III can be removed selectively to enable that amino group to be built on further. Alternatively, if the group **X** can be removed selectively, the carboxyl group can be activated for the next coupling.

The range of protecting groups is wide, and by and large conditions have been found under which they can readily be attached and later removed without causing epimerisation at the chiral centres. The reaction which causes most trouble is the sequence in which the carboxylic acid group is activated and then combined with the amino group of the other component. The chiral centre at peril in this sequence is the one adjacent to the activated carboxyl function. The reaction which is most often the cause of epimerisation at this centre is oxazolone (azlactone) formation, following the activation step. The danger stems from the presence in the protected and activated amino acid or peptide (IV) of an internal nucleophilic centre five atoms away:

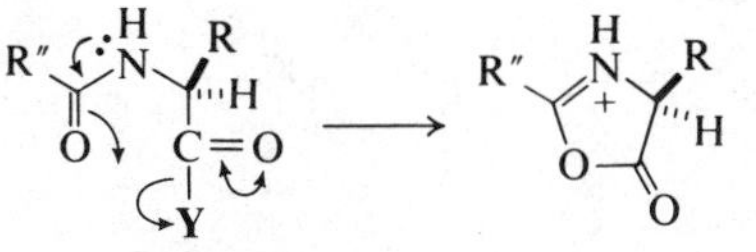

(IV)   (V)

With either acid or base, or both, the intermediate (V) can readily lose the hydrogen atom next to the carbonyl group in an enolisation which is assisted by the development of an oxazole system:

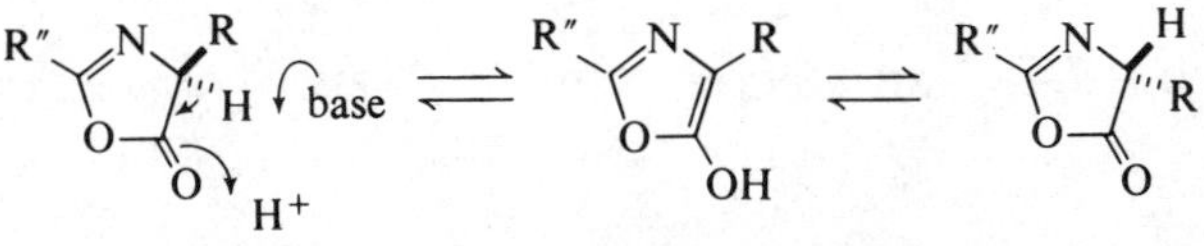

This process will lead to epimerisation. The groups R, R″, and **Y** in IV affect the ease of this process and so do the conditions under which the reaction proceeds; thus non-polar solvents, a minimum of base, and low temperature are usually desirable to avoid oxazolone formation. Amongst *N*-protecting groups (R″ in IV), the urethane type, such as the benzyloxycarbonyl (R″ = PhCH$_2$O), is particularly successful at limiting epimerisation; but this only applies, of course, to the making of the first dipeptide. After that, the protecting group is too far removed from the site where epimerisation takes place. Amongst coupling procedures, the use of the azide (IV, $Y = N_3$) is notably trouble-free, and so are some of the most recent coupling reagents, such as *N*-ethyl-5-phenyl-isoxazolium-3′-sulphonate (Woodward's reagent K) and 2-ethoxy-1-ethoxycarbonyl-1,2-dihydroquinoline (EEDQ).

All these choices of coupling technique, reaction conditions and protecting groups constitute the *tactics* of peptide synthesis. They are influenced, at least partly, by the overall *strategy;* and a few words are needed about this aspect of peptide synthesis. One strategy is linear elongation, starting at one end and adding amino acids one at a time. In the other strategy, small peptides are joined together, and the resulting oligopeptides are then themselves joined together, and so on. The latter approach is known as a *convergent* synthesis and the former as a *linear* synthesis.

Linear elongation starting with the *N*-protected amino acid has the advantage that fewer steps are needed than when starting from the other end: the acid has to be activated and can then be mixed with the free amino acid (in which the amino group is a much better nucleophile than the carboxylate group). The alternative

$$\underset{\text{ZHN}}{\overset{\text{R}\quad\text{H}}{\diagdown}}\text{CO}_2\text{H} \longrightarrow \underset{\text{ZHN}}{\overset{\text{R}\quad\text{H}}{\diagdown}}\text{COY} \xrightarrow{\underset{\text{H}_2\text{N}}{\overset{\text{R}'\quad\text{H}}{\diagdown}}\text{CO}_2\text{H}} \underset{\text{ZHN}}{\overset{\text{R}\quad\text{H}}{\diagdown}}\text{CONH}\underset{}{\overset{\text{R}'\quad\text{H}}{\diagdown}}\text{CO}_2\text{H}$$

$$\downarrow$$

$$\text{etc.} \longleftarrow \underset{\text{ZHN}}{\overset{\text{R}\quad\text{H}}{\diagdown}}\text{CONH}\underset{}{\overset{\text{R}'\quad\text{H}}{\diagdown}}\text{COY}$$

linear route, starting with the carboxy-protected amino acid and mixing it with an amino-protected, activated acid, requires two extra steps for each peptide bond: the protection of the amino group (in VI), and the removal of that protecting group (from VII) before the next combination step. However, this route has the

$$\underset{\text{ZHN}}{\overset{\text{R}\quad\text{H}}{\diagdown}}\text{CO}_2\text{H} \longrightarrow \underset{\text{ZHN}}{\overset{\text{R}\quad\text{H}}{\diagdown}}\text{COY} \xrightarrow{\underset{\text{H}_2\text{N}}{\overset{\text{R}'\quad\text{H}}{\diagdown}}\text{COX}} \underset{\text{ZHN}}{\overset{\text{R}\quad\text{H}}{\diagdown}}\text{CONH}\underset{}{\overset{\text{R}'\quad\text{H}}{\diagdown}}\text{COX}$$
$$\quad(\text{VI})\qquad\qquad\qquad\qquad\qquad\qquad\qquad\qquad(\text{VII})$$

$$\downarrow$$

$$\text{etc.} \longleftarrow \underset{\text{H}_2\text{N}}{\overset{\text{R}\quad\text{H}}{\diagdown}}\text{CONH}\underset{}{\overset{\text{R}'\quad\text{H}}{\diagdown}}\text{COX}$$

considerable advantage that it is comparatively easy to choose a protecting group, **Z**, which will minimise epimerisation, whereas less control is possible in the former route because the rest of the peptide chain is necessarily joined to the site vulnerable to epimerisation.

Both of these linear approaches suffer severely from the arithmetic demon. The overall yield is simply the product of the yields of all the steps. One of the advantages of the strategy involving the joining of small peptides is that the overall yield is not the simple product of all the yields. We can illustrate the effect on the yield of this convergent approach with an imaginary sequence as shown in Figure 1.

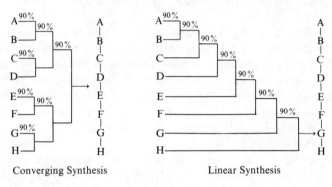

Figure 1

If the yield is 90% in all steps, the overall yield in the converging synthesis is $0.9^3 \times 100$, which equals 73%, whereas the yield in the linear synthesis is $0.9^7 \times 100$, which equals 48%. The advantage is even more striking with large polypeptides: a converging synthesis of 64 units with 90% yield at each step could give an overall yield of 53%, whereas the linear synthesis would give an overall yield of only 0.13%. With a small drop in the average yield, to 85% instead of 90%, a converging synthesis could still give a yield of 37%, but now the linear synthesis will give only 0.004%. It is obvious that, for large polypeptides, linear syntheses are impractical unless very high yields can be maintained.

Another advantage of convergent synthesis is that at each step the change from starting materials (a tetrapeptide and a pentapeptide, say) to product (a nonapeptide) is comparatively large, and consequently the separation of product from any unchanged starting materials is relatively easy.

However, the convergent approach to peptide synthesis suffers from the same disadvantage as the linear approach starting with an *N*-protected amino acid: the activated carboxyl function is necessarily part of a peptide chain, and the carbon atom adjacent to that carboxyl function is therefore exposed to the risk of epimerisation. Thus part of the success of a convergent strategy in a peptide synthesis is going to depend upon a correct choice of places at which to join fragments. Since glycine does not possess a chiral centre, it is an obvious choice for the *C*-terminal of a fragment. Proline, with an NH group instead of the usual NH$_2$ group, is also not subject to the oxazolidone pathway for epimerisation, and is another safe choice for a *C*-terminal fragment. The synthetic chemist can draw upon a large store of accumulated experience in deciding a strategy. Sometimes the reasons for a particular strategy are obvious; sometimes the eventual choice has been made only after experience of the particular compounds in question. We shall now look at the synthesis of the simple nonapeptide, bradykinin (**1**), which has been synthesised many times. This was the first synthesis.[2]

Let us look first at each of the amino acids which were used; in each case some protection of those functional groups not immediately needed was first applied. The guanidine group of arginine (**2**) was protected as its nitro derivative, and the amino group, which became the *N*-terminal of the peptide, was protected as its benzyloxycarbonyl derivative (**11**). This very common protecting group is conventionally written as Z- in abbreviated reaction schemes, such as that used for this synthesis on p. 106. The full structure of all intermediates is drawn out in the elaborate scheme on pp. 103–105.

The next seven amino acids were each protected at one group, leaving the other free: proline (**3**) as its methyl ester (**12**), proline (**4**) as its benzyloxycarbonyl derivative (**13**), glycine (**5**) as its ethyl ester (**14**), phenylalanine (**6**) as its benzyloxycarbonyl derivative (**15**), serine (**7**) as its methyl ester (**16**), proline (**8**) as its benzyloxycarbonyl derivative (**17**) and phenylalanine (**9**) as its benzyloxycarbonyl derivative (**18**). At the future *C*-terminal end, arginine (**10**) was again doubly protected, this time as its nitro derivative and *p*-nitrobenzyl ester (**19**).

The protected amino acids were next joined in pairs. The arginine derivative (**11**) and the proline derivative (**12**) were joined, using dicyclohexylcarbodiimide (DCC), see pp. 84 and 96. Because the *N*-protecting group, namely the benzyloxycarbonyl group, is of the urethane type, there is no risk of epimerisation. The proline and glycine derivatives (**13** and **14**) were combined using a different coupling procedure, introduced by Th. Wieland and known as the mixed anhydride method. The proline derivative (**13**) was treated with ethyl chloroformate and triethylamine to give the mixed anhydride (VIII). When this was mixed with the glycine ester (**14**), the amino group of the latter attacked the proline carbonyl group of the anhydride (as in VIII) rather than the other carbonyl group (as in IX),

(VIII)   (IX)

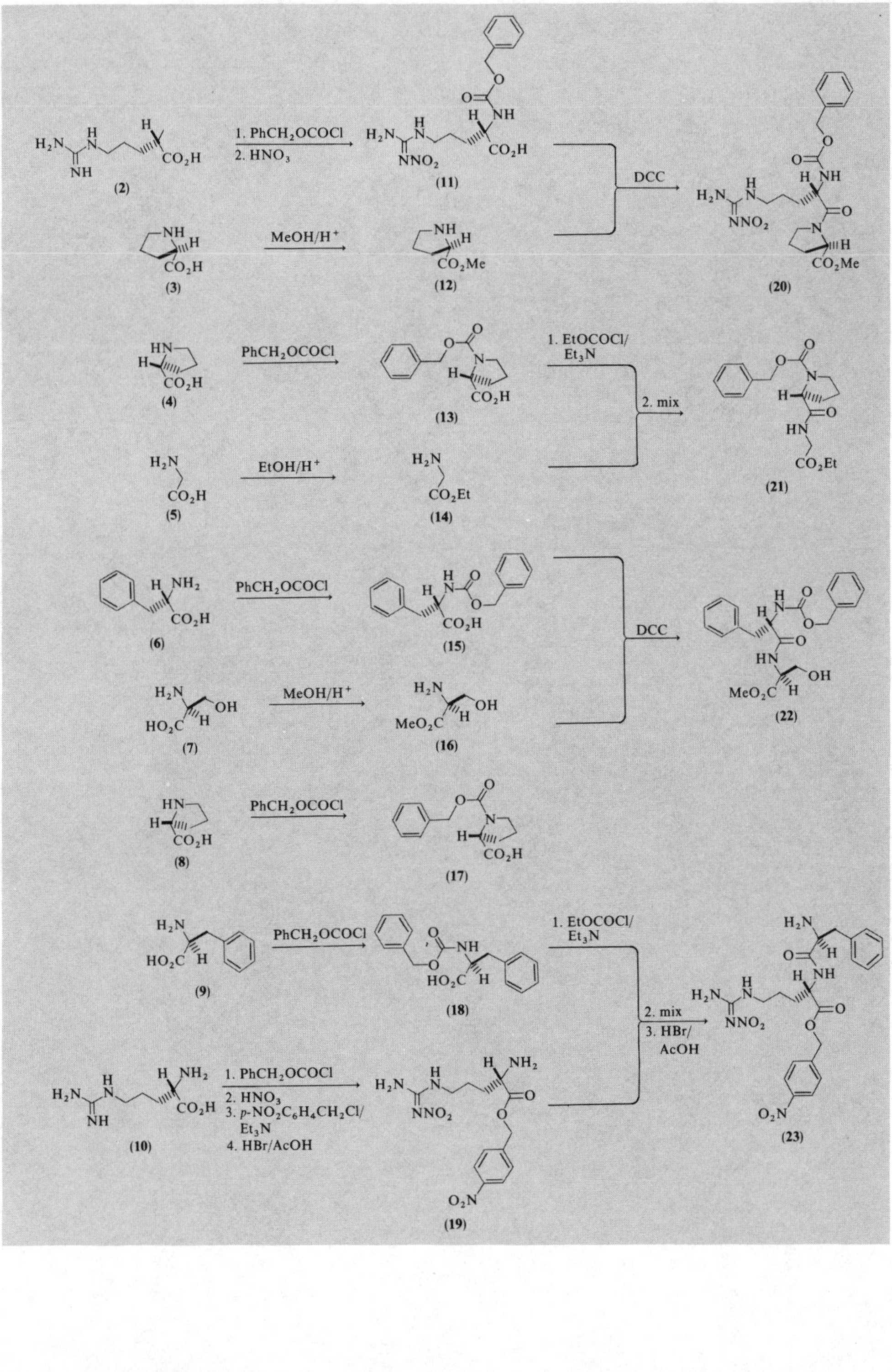
(2)
1. $PhCH_2OCOCl$
2. HNO_3
(11)
(3)
$MeOH/H^+$
(12)
DCC
(20)
(4)
$PhCH_2OCOCl$
(13)
1. $EtOCOCl/Et_3N$
2. mix
(21)
(5)
$EtOH/H^+$
(14)
(6)
$PhCH_2OCOCl$
(15)
(7)
$MeOH/H^+$
(16)
DCC
(22)
(8)
$PhCH_2OCOCl$
(17)
(9)
$PhCH_2OCOCl$
(18)
1. $EtOCOCl/Et_3N$
2. mix
3. $HBr/AcOH$
(23)
(10)
1. $PhCH_2OCOCl$
2. HNO_3
3. $p\text{-}NO_2C_6H_4CH_2Cl/Et_3N$
4. $HBr/AcOH$
(19)

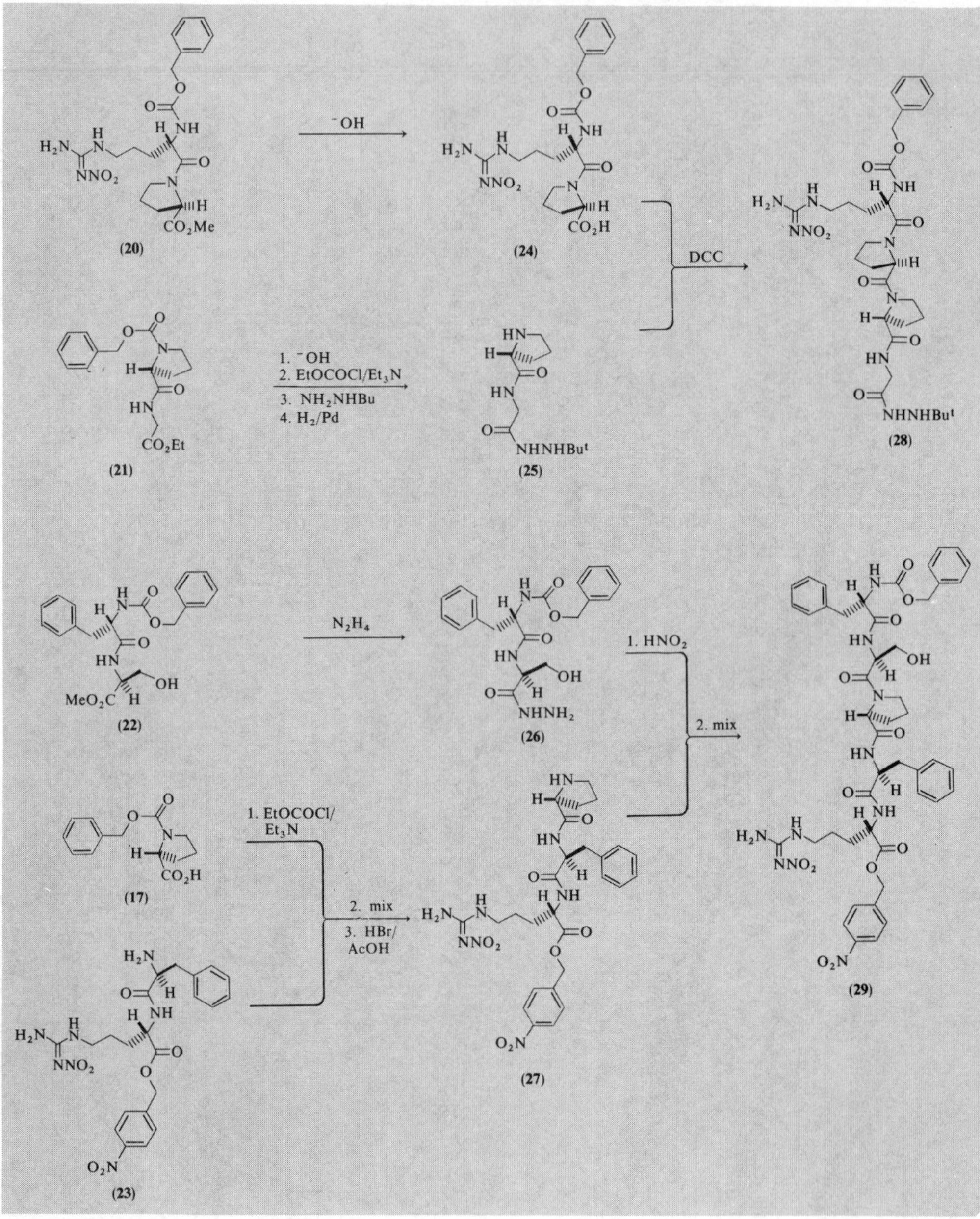

(20)

(24)

(28)

$^-$OH

DCC

(21)

1. $^-$OH
2. EtOCOCl/Et$_3$N
3. NH$_2$NHBu
4. H$_2$/Pd

(25)

(22)

(26)

(29)

N$_2$H$_4$

1. HNO$_2$

2. mix

(17)

1. EtOCOCl/
Et$_3$N

2. mix
3. HBr/
AcOH

(27)

(23)

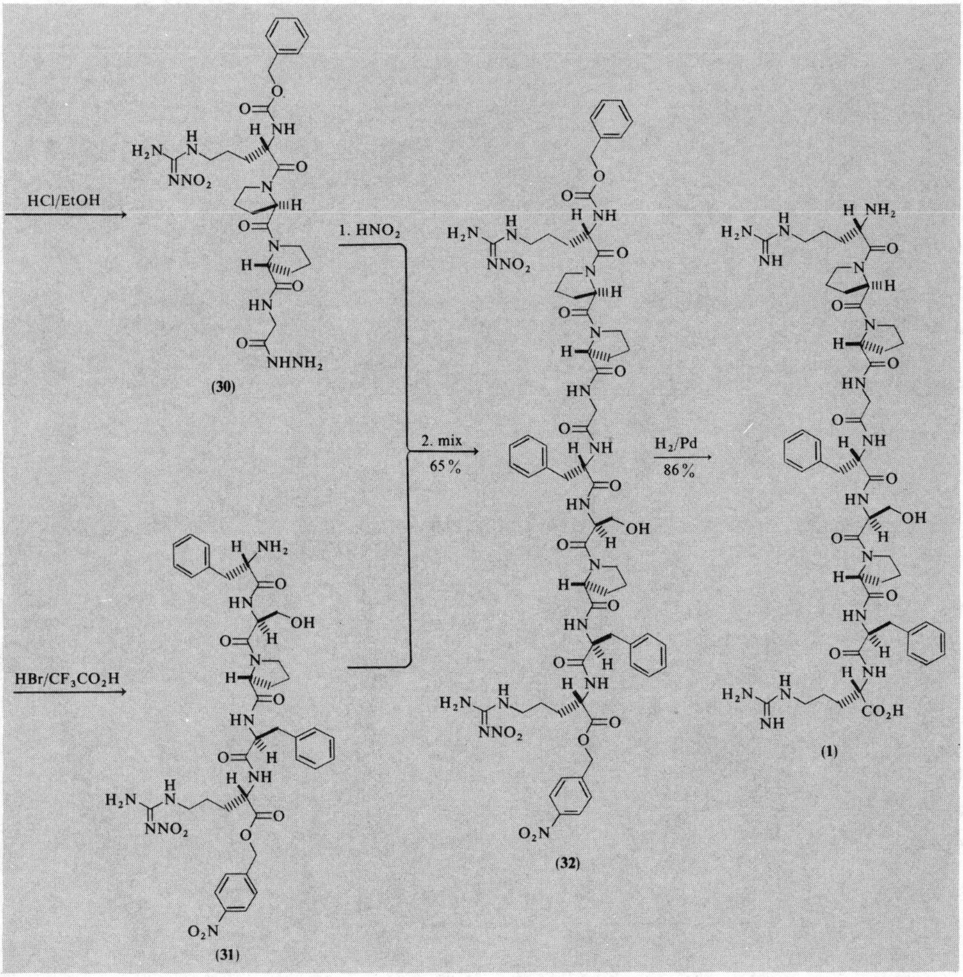
HCl/EtOH
(30)
1. HNO$_2$
2. mix
65%
(32)
H_2/Pd
86%
(1)
HBr/CF$_3$CO$_2$H
(31)

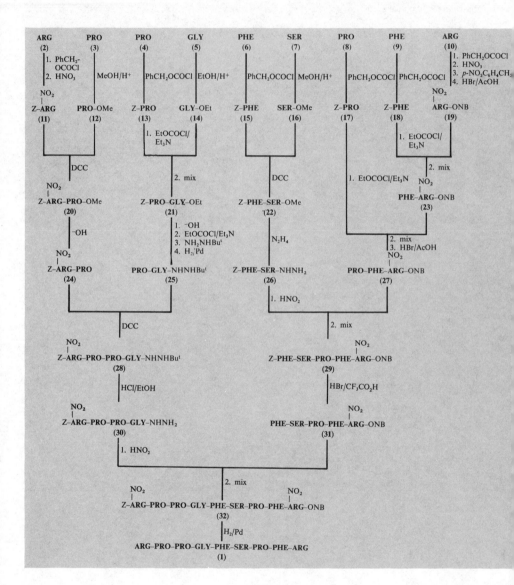

and gave the protected dipeptide (**21**). The phenylalanine and serine derivatives (**15** and **16**) were combined by the DCC method to give the protected dipeptide (**22**). The last three protected peptides were joined, by the mixed anhydride procedure. Because the coupling started at the *C*-terminal end (arginine), the activated acid function at both steps was a benzyloxycarbonyl derivative and hence not subject to racemisation. The treatment with hydrobromic acid in acetic acid selectively removed the benzyloxycarbonyl group; the more usual hydrogenolysis is not applicable in this case, because there are two other susceptible functions: the *p*-nitrobenzyl ester group and the nitro group on the arginine. At this stage in the synthesis, there are three dipeptide derivatives (**20**, **21** and **22**) and a tripeptide derivative (**27**).

The dipeptides were then modified in preparation for the next set of couplings. The ester group of **20** was removed with alkali to give the free acid (**24**). The ester group of **21** was removed and replaced by a t-butylhydrazine grouping, which will be needed at a later stage. The mixed anhydride procedure was used for this operation. It is obviously wise to do this kind of change at an early stage in the synthesis, rather than try to perform it on the more precious and more heavily functionalised tetrapeptide derivative. With the carboxyl group protected, hydrogenolysis released the amino group to give a dipeptide (**25**) ready for the next stage. The ester group of **22** was replaced directly by hydrazine to give the hydrazide (**26**).

So far, no risk of epimerisation had been run because the activated acid function at each coupling was on an amino acid protected by a benzyloxycarbonyl group. But, as the pairs of amino acids were joined, this would no longer be true. The joining of the first two pairs (**24** and **25**) would not be troublesome, because the amino acid at the *C*-terminal end (proline) was one of the two (the other being glycine) which gives no racemisation by any of the more usual methods. The DCC method was therefore quite adequate for this step, which gave a protected tetrapeptide (**28**). Joining the protected dipeptide (**26**) to the protected tripeptide (**27**) was not so straightforward: the carboxyl group involved in the coupling was that of a serine residue, and serine is one of the amino acids most susceptible to epimerisation using the DCC or mixed anhydride methods. For coupling, then, the azide method, which causes no racemisation, was used. That was the reason for the presence of the hydrazine group: with nitrous acid the hydrazide (**26**) gave the corresponding azide; then, on mixing with **27**, the amino group of the proline residue on the latter displaced the azide group, giving the pentapeptide derivative (**29**).

The protected tetrapeptide (**28**) was prepared for the last coupling by removing the t-butyl group with acid to give the hydrazide (**30**). The protected pentapeptide (**29**) was likewise prepared for the last coupling by removing the benzyloxcarbonyl group using hydrobromic acid in trifluoroacetic acid. This solvent was used because in the more usual solvent, acetic acid, the serine hydroxyl group is partially acetylated. Trifluoroacetic acid is less nucleophilic, and esterification is correspondingly slower. In many peptide syntheses, the free hydroxyl group of serine (and the free functional groups of other amino acids, such as glutamic acid, lysine, etc.) would have to be protected, as indeed the guanidino group of arginine is in this one. Fortunately, in this rather simple example, it was easy, by an expedient change of solvent, to avoid the one small problem which the presence of serine introduced.

The final coupling of the two protected peptides (**30** and **31**) was again done using the azide procedure, although the C-terminal end in this case (glycine) was not one which can epimerise. The hydrazide group was diazotised and the azide displaced by the free amino group of **31**. The protecting groups on the resultant nonapeptide derivative (**32**) were now groups all of which could be removed by hydrogenolysis. The resulting nonapeptide (**1**) was identical in physical and physiological properties to the natural product.

In recent years, a new method of peptide synthesis, introduced by R. B. Merrifield, has proved to be immensely powerful. This is a method called *solid phase* peptide synthesis, in which the C-terminal amino acid is joined on to a polymer and then extended, one unit at a time, until the polypeptide is completed and can be released from the solid support. No purification is possible, except that the removal of unused reagents, simply by filtration, is very fast and efficient. It is essential that every step give a very high yield, because this approach is necessarily that of a linear synthesis. If high yields are not achieved, the final peptide will be contaminated by peptides with one or more missing amino acid residues; purification at this stage will be difficult or impossible. Bradykinin was the first natural peptide to be synthesised by this method. The procedure which was used[3] will illustrate the technique.

The polymer used was a commercial polymer of styrene containing 2% divinylbenzene as a cross linking agent. This polymer, in the form of 200–400 mesh beads, has an open gel structure into which solvents and reagents penetrate rapidly. It was functionalised by chloromethylation:

Each of the amino acids was protected on nitrogen with the acid-labile t-butyloxycarbonyl (BOC-) group. The first to be joined to the polymer was the future C-terminal, arginine, which was additionally protected as its nitro derivative (**33**). An excess of the triethylammonium salt of this residue was stirred in ethanol with the chloromethylated polymer; under these conditions displacement (X) took place at the benzyl position and the amino acid became joined to the polymer by its carboxyl group (**34**). When this reaction was complete, the polymer, bearing

$$(X) \qquad (XI) \equiv (34)$$

the arginine residue, was simply filtered off from the excess of BOC-nitroarginine (**33**). The sequence of steps then involved: (i) the removal of the *N*-protecting group with mild acid treatment, (ii) washing with triethylamine to give the free amino group of **35** from its hydrochloride salt, (iii) transfer of the polymer to a dimethylformamide solution of the next amino acid derivative (**36**) and coupling with DCC for two hours, and (iv) washing of the polymer with dimethylformamide, ethanol, and acetic acid, to give the extended derivative (**37**). There was no risk of epimerisation, because the activated acid was protected on nitrogen as a urethane derivative. Repetition of this sequence, with appropriate *N*-protected amino acids, gave the protected nonapeptide (**38**).

With the chain complete, the polymer must be removed from the *C*-terminal end. Hydrobromic acid in trifluoroacetic acid proved suitable, just as it had for the removal of benzyl groups in the earlier synthesis. Finally catalytic hydrogenolysis of the nitro groups gave the free nonapeptide (**1**). Purification by chromatography gave bradykinin identical in physical and physiological properties to the natural material. The effective yield was 32 per cent (based on **34**), which compares favourably with 22 per cent (based on **19**) in the earlier synthesis. The solid phase synthesis is both faster and very much simpler in terms of the manipulation of intermediates. The method suffers inevitably from the disadvantages of a linear synthesis, but, because yields are very high at each step, it competes successfully with the conventional non-linear syntheses.

A dramatic illustration of the power of the solid phase method of peptide synthesis is the subsequent synthesis[4] of the enzyme ribonuclease, which contains 124 amino acids. The synthesis involved 369 chemical reactions and 11,931 operations, which were performed by an automated machine. The product had 13–24% of the activity of the natural material.

Bu^{t}OCONH   CO$_2$H

$$\equiv \begin{array}{c} NO_2 \\ | \\ BOC\text{-}ARG \end{array}$$

$$ClCH_2\text{-}\bigcirc\text{-}\textcircled{P}$$

$$\downarrow \begin{array}{l} Et_3N/BOC\text{-}ARG/EtOH \\ (33) \end{array} \quad \begin{array}{c} NO_2 \\ | \end{array}$$

HN

$=$NNO$_2$

H$_2$N   **(33)**

$$BOC\text{-}ARG\text{-}OCH_2\text{-}\bigcirc\text{-}\textcircled{P}$$

(34)

$$\downarrow \begin{array}{l} 1.\ HCl/AcOH \\ 2.\ Et_3N \end{array}$$

$$\begin{array}{c} NO_2 \\ | \\ ARG\text{-}OCH_2\text{-}\bigcirc\text{-}\textcircled{P} \end{array}$$

(35)

Bu^{t}OCONH   CO$_2$H

$$\equiv BOC\text{-}PHE$$

$$\downarrow \begin{array}{l} BOC\text{-}PHE/DCC/DMF \\ (36) \end{array}$$

(36)

$$\begin{array}{c} NO_2 \\ | \\ BOC\text{-}PHE\text{-}ARG\text{-}OCH_2\text{-}\bigcirc\text{-}\textcircled{P} \end{array}$$

(37)

$$\downarrow \begin{array}{l} 1.\ HCl/AcOH \\ 2.\ Et_3N \end{array}$$

$$\begin{array}{c} NO_2 \\ | \\ PHE\text{-}ARG\text{-}OCH_2\text{-}\bigcirc\text{-}\textcircled{P} \end{array}$$

$$\downarrow$$
$$\downarrow\ etc.$$
$$\downarrow$$

$$\begin{array}{cc} NO_2 & \quad\quad\quad\quad\quad\quad\quad\quad\quad\quad NO_2 \\ | & \quad\quad\quad\quad\quad\quad\quad\quad\quad\quad | \end{array}$$

$$BOC\text{-}ARG\text{-}PRO\text{-}PRO\text{-}GLY\text{-}PHE\text{-}SER\text{-}PRO\text{-}PHE\text{-}ARG\text{-}OCH_2\text{-}\bigcirc\text{-}\textcircled{P}$$

(38)

$$\downarrow \begin{array}{l} 1.\ HBr/CF_3CO_2H \\ 2.\ H_2/Pd \end{array}$$

$$ARG\text{-}PRO\text{-}PRO\text{-}GLY\text{-}PHE\text{-}SER\text{-}PRO\text{-}PHE\text{-}ARG$$

(1)

An equally remarkable achievement was the conventional synthesis of ribo-nuclease, carried out by R. Hirschmann and twenty-two co-workers, which gave product with comparable (if anything, lower) activity. With syntheses of this length, purity of this order of magnitude, as measured by biological activity, should be regarded as high.

These examples of protein synthesis illustrate the great advances which have have been made since the early work of Emil Fischer and, later, of M. Bergmann. The limitless number of structures and the extraordinary range of physiological activity shown by proteins and peptides make these compounds very valuable, both at present and in their potential, to medicine, physiology, and chemistry.

Questions

1. The mixed anhydride used to join phenylalanine to arginine has two electro-philic carbonyl groups, as shown in VIII and IX, one of which is more reactive than the other. Why is this?
2. Why is it necessary to protect the guanidino groups of the arginine residues, and how is nitration able to fulfil this function?
3. In the solid phase synthesis, the yield in the coupling step (**35** + **36** + DCC → **37**) is high only when an excess of DCC and of the amino acid derivative **36** is used. With less than about a two-fold excess of **36** present, a by-product (XII) is formed and some of the amino groups on the arginine residues in **35** remain unacylated.

(XII)

How did this by-product come to be formed, and how does an excess of **36** and DCC ensure the complete acylation of **35**?

References

1. M. Bodanszky and M. A. Ondetti, *Peptide Synthesis*, Interscience, New York, 1966.
2. R. A. Boissonnas, St. Guttmann and P.-A. Jaquenoud, *Helv. Chim. Acta*, **43**, 1349 (1960).
3. R. B. Merrifield, *J. Amer. Chem. Soc.*, **86**, 304 (1964); *Biochemistry*, **3**, 1385 (1964).
4. B. Gutte and R. B. Merrifield, *J. Amer. Chem. Soc.*, **91**, 501 (1969).
5. R. G. Denkewalter, D. F. Veber, F. W. Holly and R. Hirschmann, *J. Amer. Chem. Soc.*, **91**, 502 (1969), and four papers immediately following this.

Chlorophyll-*a*

(1)

The total synthesis of chlorophyll-*a* (**1**) by Woodward[1,2] in 1960 is one of the outstanding feats of synthetic chemistry. It is most remarkable for the studied boldness of the synthetic plan.

The aim was to synthesise a known derivative of chlorophyll-*a*, namely chlorin-e_6 (I = **21**), because this derivative had already been converted back to chlorophyll by Dieckmann cyclisation (I → II), ester exchange (II → III), and insertion of magnesium (III → **1**). However, chlorin-e_6 trimethyl ester (I) was a considerably more challenging synthetic objective than were the porphyrins involved in the synthesis of haemin, described in an earlier chapter (p. 40). The principal challenge was the presence of the two extra hydrogens in ring D. Dihydroporphyrins, with the two extra hydrogens at C-7 and C-8, are called, as a class, *chlorins;* the word *porphyrin* is reserved for the class of compounds at the next higher oxidation level, with a double bond between C-7 and C-8. In spite of the problem of the two extra hydrogens, Woodward chose to begin by ignoring them. He began by preparing, in the first instance, a *porphyrin* (IV).

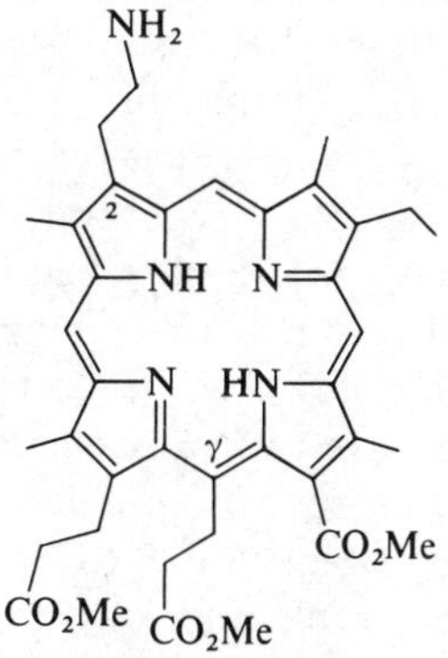

(IV)

His grounds for leaving so important a task as the introduction of the extra hydrogens to a late stage in the synthesis are important. As a result of a thorough and critical reading of the literature of chlorophyll chemistry, Woodward deduced that a number of unexplained features in the reactions shown by chlorophyll derivatives could be a consequence of an unusual degree of steric crowding on the

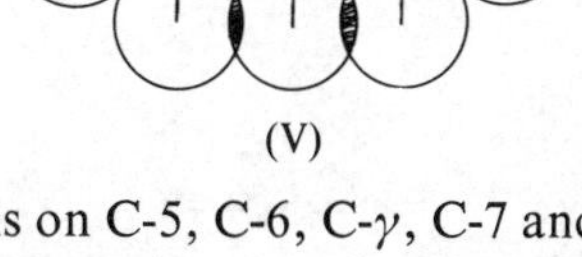

(V)

lower periphery (V), that is on C-5, C-6, C-γ, C-7 and C-8. For example, dehydro-genation of chlorins is in general easy only when there is no substituent on C-γ. When there is a substituent on C-γ, as in chlorophyll itself, dehydrogenation is considerably more difficult, because, in the course of such a reaction, the substit-uents on C-7 and C-8, which are initially above and below the plane of the ring, must be moved into the plane of the ring (VI → VII) with a consequent increase in crowding.

(VI)　　　　　　　　　　　　　　　(VII)

If steric crowding is, indeed, the explanation, then there was some hope that a properly constituted porphyrin, to be derived from IV, might be induced to transfer hydrogen atoms to the trigonal atoms at C-7 and C-8 and hence to relieve the steric compression present in the porphyrin.

With this plan in mind, it was essential that the porphyrin (IV) be made on a large scale. Not only was it inevitable that several stages of the synthesis would have to be carried out after the porphyrin stage had been reached, but it was also obvious that those stages would need much searching out. There was no sequence of well-precedented reactions to follow in order to transfer the hydrogens to C-7 and C-8. Such reactions would have to be discovered by studying the chemistry of IV and, probably, of a large number of compounds derived from it, the features of which might be supposed to be favourable to hydrogen migration. A general synthesis, giving good yields of porphyrins, was not known at the time and so the next part of the plan to be tackled was the development of a porphyrin synthesis which would give good yields, the older methods used by Hans Fischer (p. 45) being plainly inadequate.

The method used for the synthesis of porphyrins[3] was based on that for making dipyrrylmethene salts, in which the combination of a pyrrole having a free α-position with a pyrrole-α-aldehyde gives good yields of the well stabilised dipyrrylmethene salts (see also p. 43):

For a porphyrin synthesis based on this method, two dipyrrylmethanes (VIII and IX) were needed, both halves of which could give dipyrrylmethene salt systems (X):

(VIII)　　　(IX)　　　　　　　　　(X)　　　　　　　　　(XI)

Oxidation of this product (X) would then give the corresponding porphyrin (XI). It was likely that this route would give high yields, because both halves of the system would be taking part in a kind of reaction known to be clean; this was not the case in the earlier porphyrin syntheses.

It remained, in planning the synthesis, to define just which porphyrin to make. The compound (IV = 11) was chosen because it possessed functional groups suitable for conversion to those of chlorin-e_6 (notably, the aminoethyl chain on C-2 is a potential vinyl group; but, unlike the vinyl group itself, it is not likely to polymerise or otherwise frustrate attempts to perform reactions elsewhere in the molecule). But this porphyrin has one more carbon atom in the chain on C-γ than is needed; it has this extra carbon atom because, without it, there is a smooth pathway (XII → XIII and arrows) for the loss of an acetic acid side chain, a loss

$$-H^+ \text{ from } c\text{-}\delta$$
$$-H^+ \text{ from } c\text{-}\beta$$
$$+H^+ \text{ on } c\text{-}\gamma$$

(XII≡X)   (XIII)   (XI, R = H)

which should be assisted by the relief of steric compression in this part of the molecule. A propionic ester chain on C-γ would not be lost by such a pathway.

All these considerations were combined and put into practice: the dipyrrylmethanes (5 and 9) were prepared, they were combined in high yield, and the product was oxidised to a porphyrin (11); further reactions of this porphyrin were found in which hydrogen transfer was indeed observed, and chlorin-e_6 trimethyl ester (21) was synthesised. We shall now look, in outline, at the individual steps of this synthesis.

In the first place, four pyrroles (2, 3, 6 and 7) were prepared by standard methods, along the lines used by Fischer in his syntheses. They were then combined in pairs, the left-hand pair to give a dipyrrylmethene salt (4), which was reduced to a dipyrrylmethane (5), and the right-hand pair directly to a dipyrrylmethane (8). The latter dipyrrylmethane was acylated, and the protecting group removed with alkali, to give the dipyrrylmethane (9).

The next stage was the combination of the two dipyrrylmethanes according to the plan (VIII + IX → X); but, in the real situation, it was not quite so simple: the dipyrrylmethanes (5 and 9) were unsymmetrical and could (in fact did) react both ways round. An ingenious solution to this problem was the combination of

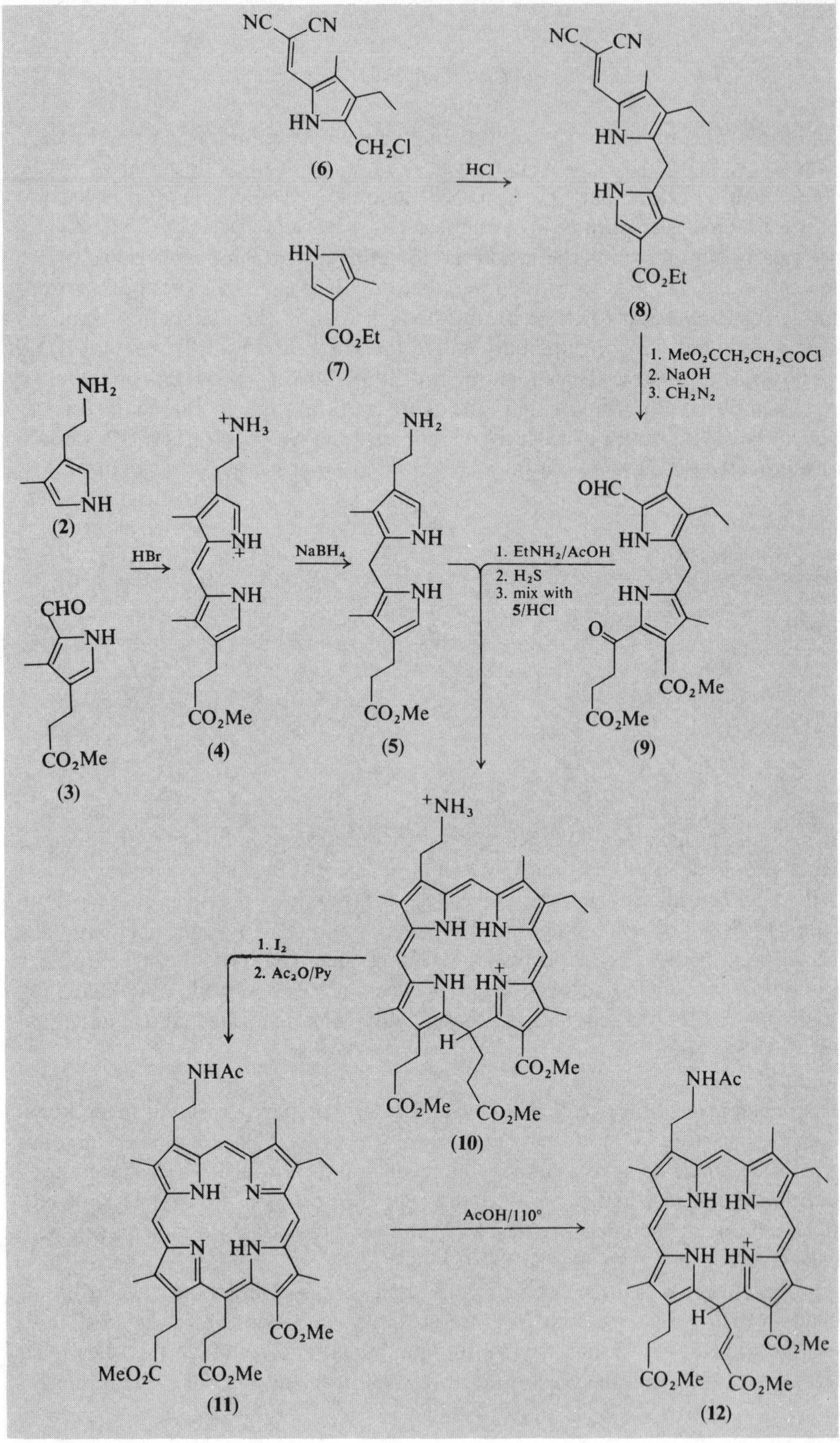
(6)
HCl
(7)
(8)
1. MeO2CCH2CH2COCl
2. NaOH
3. CH2N2
(2)
(3)
HBr
(4)
NaBH4
(5)
1. EtNH2/AcOH
2. H2S
3. mix with
5/HCl
(9)
(10)
1. I2
2. Ac2O/Py
(11)
AcOH/110°
(12)

the amino group of **5** with the formyl group of **9** to give a Schiff base (XIV). This derivative now held the nucleophilic and electrophilic carbon atoms in close proximity, and guaranteed the desired orientation. However, this idea was more

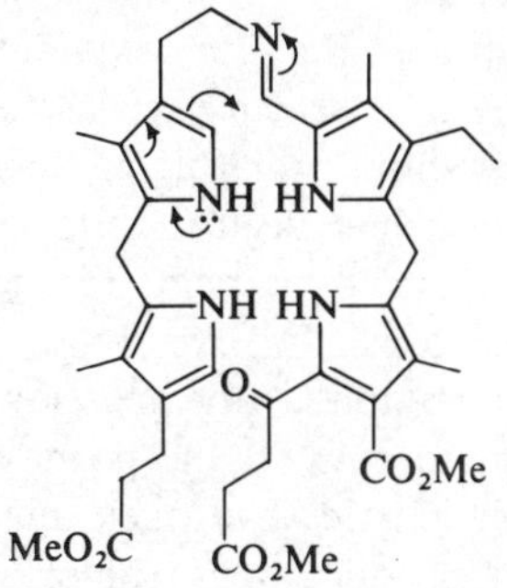

(XIV)

difficult to put into practice than it might seem. Pyrrole α-aldehydes are vinylogous amides and are not, therefore, as electrophilic as ordinary aldehydes. Furthermore, dipyrrylmethanes, such as **5**, bearing only alkyl groups, are, like pyrroles and unlike dipyrrylmethenes, sensitive and unstable compounds, particularly in the presence of acid. The result was that no conditions could be found for the direct formation of the Schiff base: acid catalysis was needed to form the Schiff base, but the acid always caused the disintegration of the dipyrrylmethane system of the left hand side. However, a Schiff base (XV) of the right hand side with ethylamine could be prepared, and this with hydrogen sulphide gave the thioaldehyde (XVI).

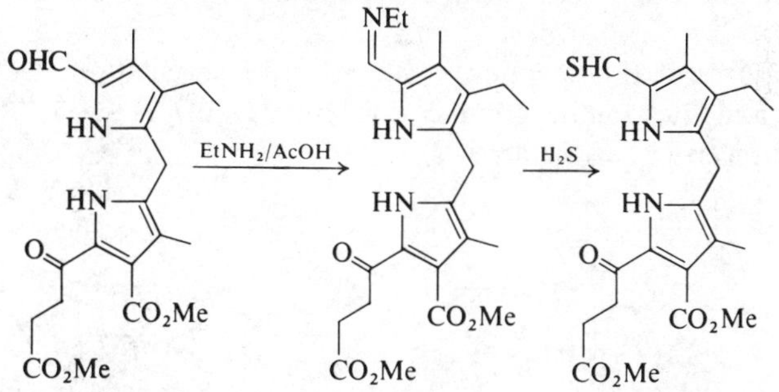

(9)   (XV)   (XVI)

The thioaldehyde, unlike the oxygen analogue, formed Schiff bases without the need for acid-catalysis, and it was possible, by mixing the thioaldehyde with the freshly prepared solution of the dipyrrylmethane (**5**), to prepare the Schiff base (XIV) and to use it for the cyclisation step. The product first isolated was the cation (**10**). The cyclisation step gave a bis-dipyrrylmethene (like X), which would probably lose a proton from the β- or δ-position; tautomerism of the resultant cation then led to the cation (**10**) (see XII → XIII). That this compound, rather than the first-formed tautomer, should have been the product is no doubt a consequence of the steric compression phenomenon. The macrocyclic conjugated system of **10** (and of XIII) is tautomeric with (i.e. at the same oxidation level as) the *chlorin* system; this system is called a *phlorin*. In practice it was found to be best to oxidise the intermediate phlorin (**10**) without isolating it and to acetylate the product to give the porphyrin (**11**), which had been the first objective of the synthesis. The yield of this porphyrin from the components (**5** and **9**) was an impressive 50 per cent after a series of five reactions.

With the porphyrin now complete, the reaction to be looked for was some kind of hydrogen migration to convert the porphyrin into a chlorin. When the porphyrin was heated in acetic acid, hydrogen migration did occur, but did not give a chlorin. Instead, the product was a phlorin salt (**12**), which was readily recognised because of the similarity of its ultraviolet-visible spectrum to that of the phlorin salt (**10**) isolated earlier. Although phlorins are tautomers of chlorins, this compound could not be induced to rearrange to a chlorin. However, oxidation gave a porphyrin (**13**), and when this porphyrin was heated in acetic acid it was converted to a chlorin (**14**). (This reaction is an equilibrium reaction, in which the cyclised material (**14**) is predominant.) Another product was the result of cyclisation to C-6, but it was only a very minor component, probably because conjugation of the methoxycarbonyl group with the ring system is interrupted in that product, whereas no such loss of conjugation occurs in the major product (**14**), and probably also because the relief of steric strain is greater when the propionic ester chain, rather than the methoxycarbonyl group, is lifted out of the plane of the ring.

The mechanism of this reaction is a matter for speculation: C-7 is obviously nucleophilic and the double bond electrophilic (XVII arrows), but these roles can easily be reversed (XVIII arrows):

MeO$_2$C   CO$_2$Me   MeO$_2$C   CO$_2$Me

(XVII)   (XVIII)

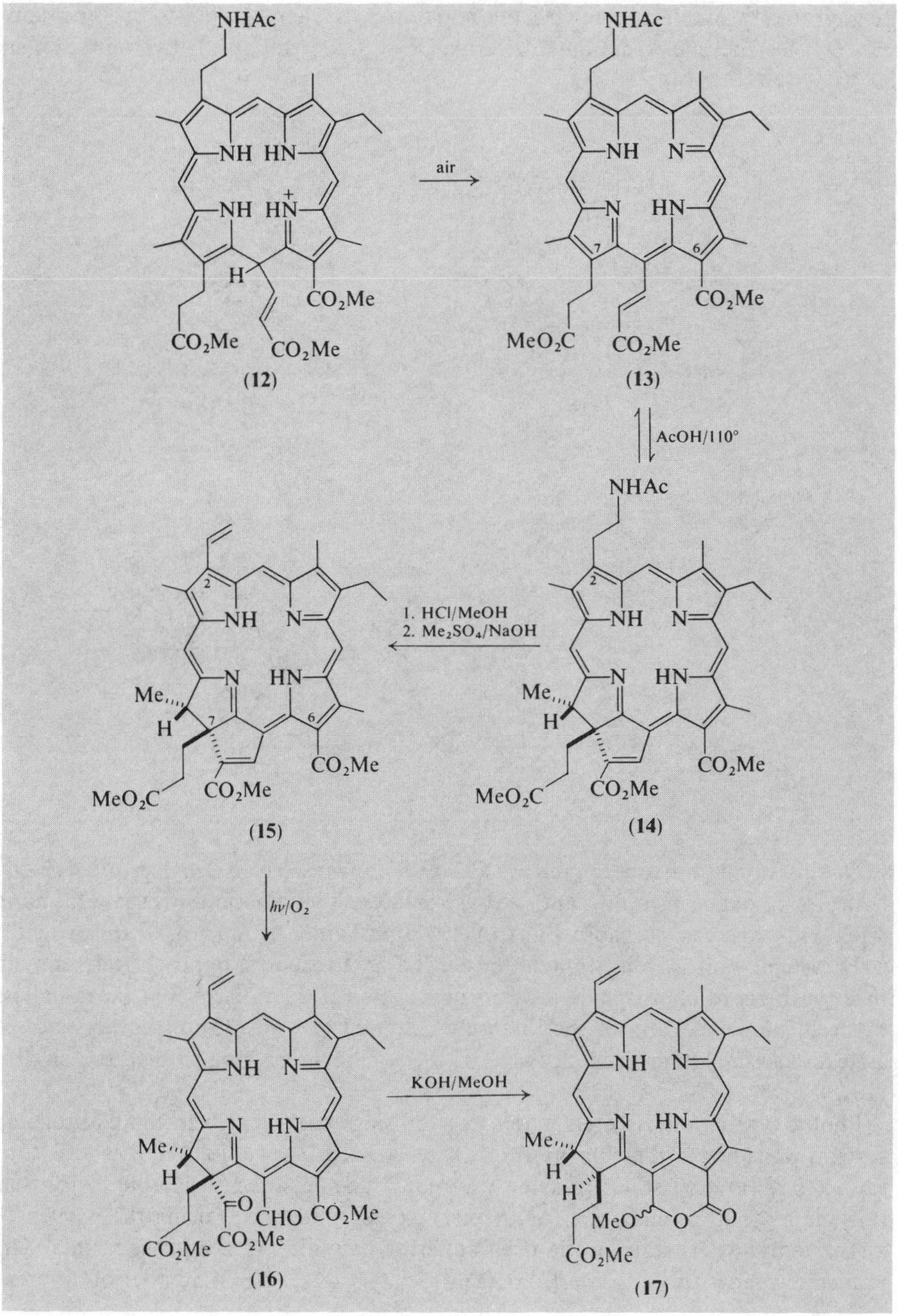
NHAc
NH HN
NH HN
H
CO_2Me
CO_2Me
CO_2Me
(12)
air
(13)
MeO_2C
AcOH/110°
(14)
1. HCl/MeOH
2. Me_2SO_4/NaOH
(15)
Me
$h\nu/O_2$
(16)
CHO
KOH/MeOH
(17)
MeO

It is probably simplest to regard the reaction as a dienyl cation-to-cyclopentenyl cation electrocyclic reaction (XIX ≡ XX → XXI) followed by tautomerism (XXI → XXII ≡ **14**):

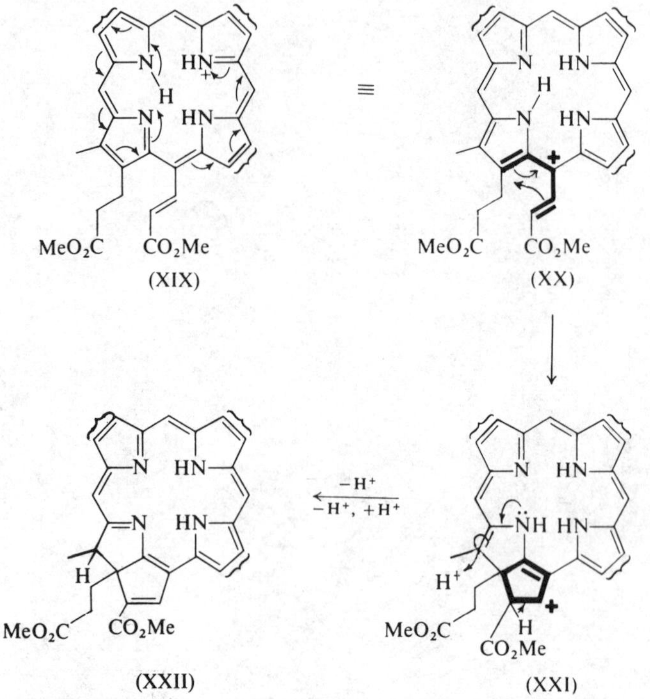

At this point, the side chain on C-2 was converted to a vinyl group (**15**) by hydrolysis, quaternisation, and a base-catalysed elimination of the Hofmann type. This was easy because the trimethylammonium group was β to an acidic C-H system, as it is in a Mannich base. The next reaction, appropriately enough in a synthesis of chlorophyll, was a photochemical one, discovered more or less by accident. Irradiation of the intensely coloured chlorin (**15**) in the presence of air caused a highly specific cleavage of the cyclopentene ring to give the chlorin (**16**).

Photooxygenation is often a reaction of singlet oxygen. The light absorbing system, presumably the chlorin in this case, absorbs the (visible) light and is thus raised to an excited state; in some way not yet understood, it is able to transfer this energy to a ground state (i.e. triplet) oxygen molecule, raising the oxygen to an excited (singlet) state, while itself suffering demotion to the ground state. The excited oxygen is then a powerful reagent, which can react in a number of different

ways. With dienes, the usual reaction is of the Diels-Alder type (e.g. XXIII →
XXIV),[4] and with olefins having an allylic hydrogen atom, the usual product is an
allylic hydroperoxide (e.g. XXV → XXVI).[5] With some olefins, apparently with

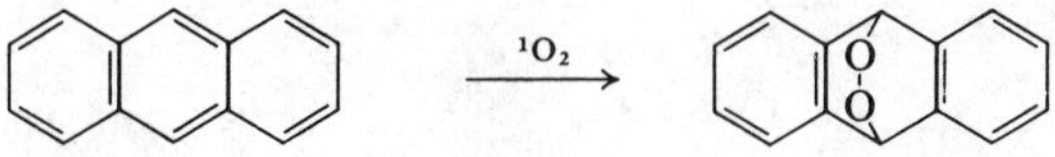

(XXIII)   (XXIV)

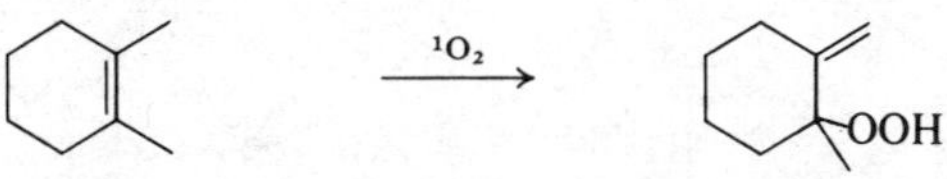

(XXV)   (XXVI)

(XXVII)   (XXVIII)   (XXIX)

this one and certainly with electron-rich olefins such as enamines, singlet oxygen
undergoes a cycloaddition reaction to give 1,2-dioxetans (XXVII → XXVIII),[6]
which open to give two carbonyl groups (XXIX). It is not, of course, known whether
this is the sequence followed in the transformation of **15** to **16**, but it is at least a
simple account of a possible route.

On treatment with sodium methoxide, the methoxalyl group on C-7 of the
chlorine (**16**) was lost because it is part of a β-dicarbonyl-like system (XXX
arrows); at the same time, the base attacked the aldehyde group (XXXI arrows)

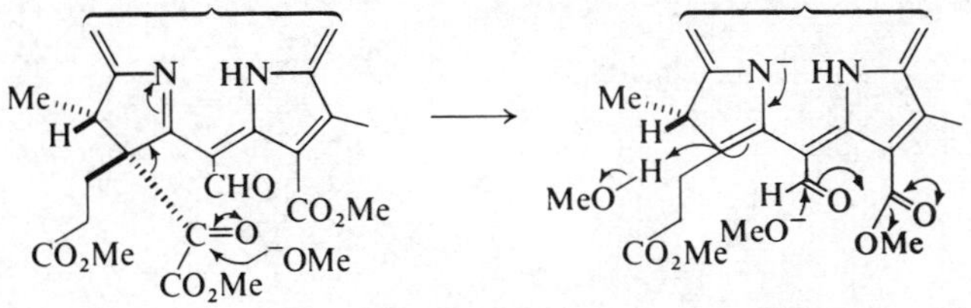

(XXX)   (XXXI)

so that the final product of this reaction was the chlorin (**17**). This compound (**17**)
now had the two extra hydrogens in place; because the last step probably allowed
equilibration at both C-7 and C-8, it is very likely that the side chains were *trans*
to one another. Furthermore, since (**17**) was the racemic form of a known compound, it was possible for the first time to compare the synthetic product with
material from natural sources.

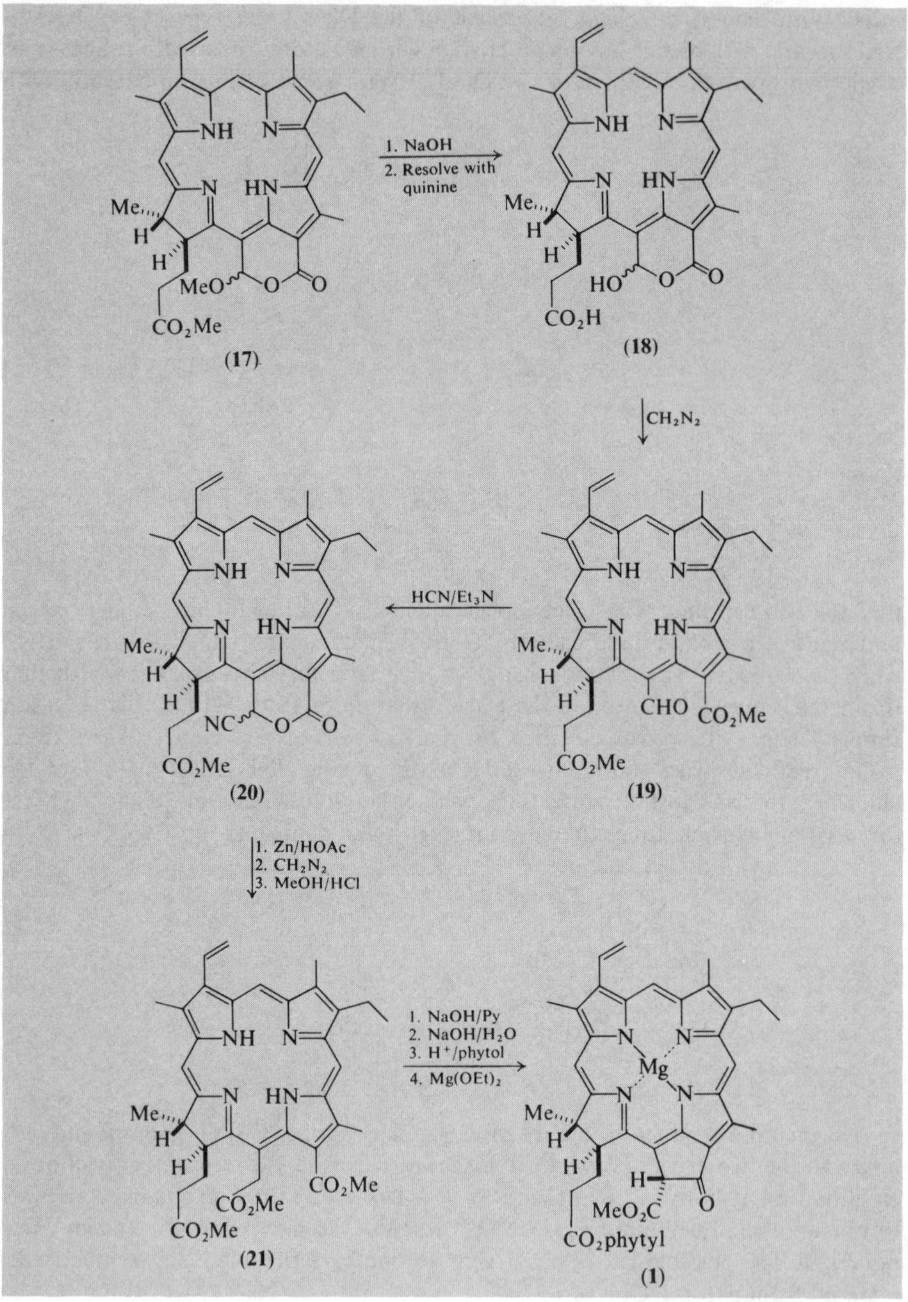

(17)

1. NaOH
2. Resolve with quinine

(18)

CH₂N₂

(19)

HCN/Et₃N

(20)

1. Zn/HOAc
2. CH₂N₂
3. MeOH/HCl

(21)

1. NaOH/Py
2. NaOH/H₂O
3. H⁺/phytol
4. Mg(OEt)₂

(1)

Hydrolysis with dilute alkali gave the acid (**18**), in which both methoxy groups have been removed. This was resolved, and then methylated with diazomethane to give the (optically active) diester (**19**), identical in every way with material from natural sources. Cyanohydrin lactone (**20**) formation, reduction, methylation, and methanolysis completed the synthesis of the triester of chlorin-e_6 (**21**). This was the ester which had already been converted by Dieckmann cyclisation to methyl phaeophorbide-*a* (II). The latter had been converted to chlorophyll-*a* itself by ester exchange with phytol and by insertion of magnesium ion. Phytol had been synthesised earlier by Burrell, Jackman and Weedon.[6]

A total synthesis does not necessarily complete the chemical study of a natural product. As Woodward said in a lecture[1] based on the work just described: '... work in chlorophyll chemistry has by no means been exhausted—indeed, we feel that our studies have opened up many more avenues than they have traversed, and we do not hesitate to hazard the opinion that the area is one from which much increase in chemical knowledge and understanding is to be had in the future'. This prediction has proved to be true: one may take the measure of its accuracy by looking at a review[7] of work done in the general field of chlorophyll chemistry after this synthesis.

Questions

1. Derive syntheses of the four pyrroles (**2**, **3**, **6** and **7**). The syntheses actually used may be found in reference 2.
2. Why is it possible to make the imine (XV) of the right hand side? In other words, why is this dipyrrylmethane capable of surviving acidic conditions, and why is only the the aldehyde group attacked?
3. In the combinations of the pyrroles (**2** with **3**, and **6** with **7**), there were two possible products in each case. Why did the pyrroles (**2** and **7**) react at the carbon atoms which gave the products observed (**4** and **8** respectively) rather than at the other free α-positions?

References

1. R. B. Woodward, *Pure and Applied Chemistry*, **2**, 383 (1961).
2. R. B. Woodward, W. A. Ayer, J. M. Beaton, F. Bickelhaupt, R. Bonnett, P. Buchschacher, G. L. Closs, H. Dutler, J. Hannah, F. P. Hauck, S. Itô, A. Langemann, E. Le Goff, W. Leimgruber, W. Lwowski, J. Sauer, Z. Valenta and H. Volz, *J. Amer. Chem. Soc.*, **82**, 3800 (1960).
3. This approach to porphyrin synthesis was developed independently by MacDonald, see G. P. Arsenault, E. Bullock and S. F. MacDonald, *J. Amer. Chem. Soc.*, **82**, 4384 (1960).

4. C. S. Foote, S. Wexler, W. Ando and R. Higgins, *J. Amer. Chem. Soc.*, **90**, 975 (1968).
5. W. Fenical, D. R. Kearns and P. Radlick, *J. Amer. Chem. Soc.*, **91**, 7771 (1969); but see also C. S. Foote, T. V. Fujimoto and Y. C. Chang, *Tetrahedron Letters*, 45 (1972).
6. P. D. Bartlett and A. P. Schaap, *J. Amer. Chem. Soc.*, **92**, 3223 (1970); S. Mazur and C. S. Foote, *J. Amer. Chem. Soc.*, **92**, 3225 (1970).
6. J. W. K. Burrell, L. M. Jackman and B. C. L. Weedon, *Proc. Chem. Soc.*, 263 (1959); J. W. K. Burrell, R. F. Garwood, L. M. Jackman, E. Oskay and B. C. L. Weedon, *J. Chem. Soc.* (*C*), 2144 (1966).
7. H. H. Inhoffen, J. W. Buchler and P. Jäger, in L. Zechmeister (Ed.), *Progress in the Chemistry of Organic Natural Products*, Vol. 26, Springer-Verlag, Vienna, 1968, p. 284; see also, H. H. Inhoffen, *Pure and Applied Chemistry*, **17**, 443 (1968).

Patchouli Alcohol

$$(1) \qquad (2) \qquad (3)$$

Patchouli alcohol (1) is the main constituent of patchouli oil, of Indian origin, an oil used in the perfume industry. The structure determination is uniquely interesting. The structure was supposed to be **2**, and *the synthesis[1] was planned and successfully executed on this basis.* Only after the synthesis was complete did an X-ray-crystallographic investigation[2] reveal that a rearrangement had taken place in both the course of the structure determination and the synthesis: the true structure was **1**.

Pyrolysis of the acetate of **1** had given patchoulene, the structure of which had been proved unambiguously to be **3**: the pyrolysis of esters generally gives an olefin by a concerted six-centre elimination pathway (I). In the case of **1**, the olefin produced by such a process would be at a bridgehead, and would be highly strained. On the other hand, the orbitals are lined up for a concerted alkyl migration to accompany the loss of acetic acid:

It was this rearrangement which had gone undetected.

The synthesis was directed first to give patchoulene (**3**), because conversion of **3** to patchouli alcohol (**1**) had already been done.

The starting material (which had been synthesised at the beginning of the century[3]) was (+)-camphor; oxidation with selenium dioxide gave camphorquinone (**4**). Diazomethane reacted with a carbonyl group in **4**, and bond migration and loss of nitrogen led to homocamphorquinone (**5**). The ethyl ether (**6**) was reduced and the resulting alcohol treated with acid to give the unsaturated ketone (**7**).

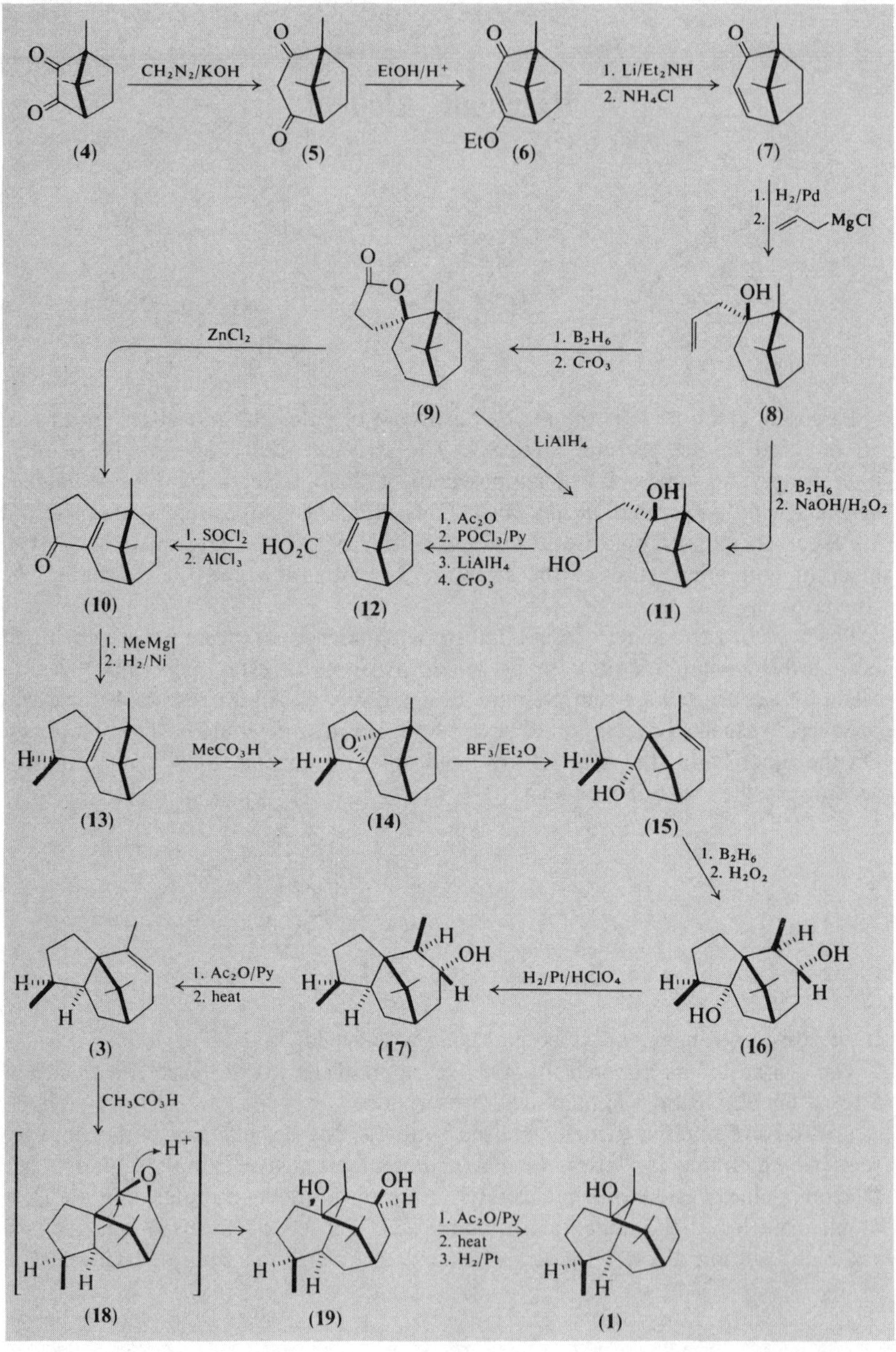
CH2N2/KOH
EtOH/H+
1. Li/Et2NH
2. NH4Cl
EtO
(4)
(5)
(6)
(7)
1. H2/Pd
2. MgCl
OH
ZnCl2
1. B2H6
2. CrO3
(9)
(8)
LiAlH4
1. B2H6
2. NaOH/H2O2
1. SOCl2
2. AlCl3
HO2C
1. Ac2O
2. POCl3/Py
3. LiAlH4
4. CrO3
HO
(10)
(12)
(11)
1. MeMgI
2. H2/Ni
MeCO3H
BF3/Et2O
(13)
(14)
(15)
1. B2H6
2. H2O2
1. Ac2O/Py
2. heat
H2/Pt/HClO4
(3)
(17)
(16)
CH3CO3H
H+
1. Ac2O/Py
2. heat
3. H2/Pt
(18)
(19)
(1)

Hydrogenation then gave homocamphor, which was treated with allyl magnesium chloride to give the alcohol (**8**), in which the allyl group has attacked from the less hindered side, opposite to the gem-dimethyl group. The chiral centre created stereoselectively in this reaction was destined to become trigonal again at a later stage. Nevertheless the stereoselectivity at this stage had the practical advantage that a single product was obtained, crystalline and with clean spectra.

The double bond of **8** reacted with diborane to give the corresponding borane, which was oxidised directly with chromic acid in acetone to give the lactone (**9**). Treatment of **9** with zinc chloride gave **10** as the major ketone. This reaction was initiated by the Lewis acid:

and the intermediate unsaturated acid (III) then underwent an intramolecular, aliphatic Friedel-Crafts reaction represented notionally as IV. However the reaction was not in fact so simple, because the ketone obtained was largely *racemic*. A more careful route from **8** or **9** to **10** was needed.

Hydroboration and alkaline hydrogen peroxide treatment of **8**, or lithium aluminium hydride reduction of **9**, gave the same diol (**11**). Treatment with, successively, acetic anhydride (giving the monoacetate), phosphorus oxychloride (giving the olefin-monoacetate), lithium aluminium hydride (giving the olefin-alcohol), and chromic acid gave the olefinic acid (**12**). The acid chloride of **12** with aluminium chloride gave **10** with full optical activity.

The reaction of **10** with methyl Grignard gave, after work up, a mixture of dienes; evidently the first-formed tertiary alcohol had suffered unusually easy dehydration. However, hydrogenation did not affect the well-embedded, tetra-substituted double bond and led to the mono-olefin (**13**), in which hydrogen had approached from the side away from the gem-dimethyl group. The structure of this product was confirmed by comparison with β-patchoulene, obtained as a minor product in the pyrolysis of patchouli acetate and as the major product of acid-catalysed dehydration of patchouli alcohol. β-Patchoulene (**13**) was epoxi-dised on the lower surface of the double bond to give **14**, which rearranged on treatment with boron trifluoride (V, arrows) to give the olefinic alcohol (**15**). Hydroboration-oxidation of the double bond in **15** gave the diol (**16**). Catalytic hydrogenation of this diol in the presence of perchloric acid gave the alcohol (**17**), either by direct reduction of the carbonium ion derived from the tertiary

(V)

alcohol, or by hydrogenation of the olefins derived from that carbonium ion. In either case the hydrogen was delivered to the less hindered side. The acetate of **17** was pyrolysed to give α-patchoulene (**3**).

Peracetic acid oxidation in a buffered medium gave an epoxide (presumably the α-epoxide corresponding to **18**) and a rearranged diol (**19**), presumably formed via the epoxide (**18**). The monoacetate of this diol (secondary alcohols are generally more rapidly acetylated than tertiary alcohols) gave, on pyrolysis, an unsaturated alcohol, which was hydrogenated to give patchouli alcohol, identical with the natural material.

The reactions described in the last paragraph were originally interpreted as giving **2**, and the existence of the rearrangement step (**18** → **19**) was not appreciated. This was not an oversight. A critical reaction, the Wolff-Kishner reduction of the ketone (VI) derived by oxidation of **19**, was said to give α-patchoulene (**3**) [as it

(VI)   (VII)

would have, had the original formulation for this ketone (VII) been correct]. The actual product of the Wolff-Kishner reduction later proved to be patchouli alcohol itself, a result entirely compatible with the structure (VI) for the ketone and (**1**) for patchouli alcohol.

Questions

1. The sequence II → III → IV → **10**, as formulated on p. 127, would not give a racemic product. If a symmetrical ion or molecule is in equilibrium with one of the intermediates likely to be involved in this sequence, then racemisation will be observed. A minor ketone, isomeric with **10** and obtained from the reaction mixture, was optically inactive; it had λ_{max} (EtOH) 247 nm (ε 14,500); ν_{max} (CCl$_4$) 1665 and 1615 cm^{-1}; τ 4·25 (1H, singlet) and τ 8·82 (6H, singlet) as well

as broad absorption in the methylene region (13H). This product has probably been produced from the symmetrical intermediate needed to explain the racemisation. Identify first the symmetrical intermediate and then the ketone.

2. Hydroboration-oxidation has not been elaborated upon in the text. If you are not familiar with the reaction, you should look it up; it is a very important synthetic tool, especially for the anti-Markownikoff hydration of double bonds.

References

1. G. Büchi, W. D. MacLeod and J. Padilla, *J. Amer. Chem. Soc.*, **86**, 4438 (1964).
2. M. Dobler, J. D. Dunitz, B. Gubler, H. P. Weber, G. Büchi and J. Padilla, *Proc. Chem. Soc.*, 383 (1963).
3. G. Komppa, *Ber.*, **36**, 4332 (1903); *Annalen*, **370**, 209 (1909).

A Steroid Synthesis on an Industrial Scale

$$CO_2CH_2OAc$$

(1)

We met in an earlier chapter (p. 57) the first synthesis of a saturated steroid. Since the time of that synthesis, many syntheses of steroids, following many different approaches, have been reported. None of them, however, was able to compete on an industrial scale with partial syntheses based on plant sources. In 1960 the first commercial total synthesis of a steroid was announced by L. Velluz;[1,2] the route used was subsequently improved. The improved route[3] takes advantage, in its early stages, of a synthesis of 1953 by the CIBA group,[4] but possesses two unique features of interest. The first is that the synthesis is *convergent:* that is, large fragments are assembled and then joined together as late as possible. We met this term in connection with peptide synthesis; as in that field, so in this: a convergent synthesis is inherently preferable to a *linear* one, both in terms of yield and in terms of the volumes of solvent and quantities of reagents needed. The second feature is that resolution is performed at a relatively early stage; this too is an advantage on an industrial scale, because the unwanted enantiomer is not carried right through the synthesis, consuming solvents and reagents, only to be thrown away at the end. These excellent features of design, the high yield, and the large range of structures which can be prepared from the product, make this route a good synthesis of steroids on a ton scale.

The unsaturated ketone (3), which was to provide the chain of atoms C-5 to C-11, was prepared[5] from glutaric anhydride (2) by a sequence involving an aliphatic Friedel-Crafts reaction on ethylene. This unsaturated ketone was joined to the ring D component, 2-methylcyclopentan-1,3-dione (4), in a Michael reaction. The product (5) was reduced microbiologically to the alcohol (6), in greater than 70 per cent yield.[6] This reduction of the prochiral* compound (5) gave only one

---

* A prochiral centre is one in which two of the substituent groups are alike (as in CH_2XY), so that a change in one of them (to CHXYZ in this case) would lead to a chiral centre. In the case of 5, reduction of one of the two carbonyl groups in the ring will make C-13 a chiral centre; thus C-13 in 5 is *prochiral*.

of the two possible diastereoisomers and only the one enantiomer. It was particularly efficient because not only did it effect a resolution early in the synthesis but also, unlike a conventional resolution, it did not discard half the starting material. An acid-catalysed aldol condensation was used to close ring C, giving the α,β-unsaturated ketone (7), in which hydrolysis of the ester function had also taken place.

Catalytic hydrogenation of 7 gave the saturated product (8), in which hydrogen had been delivered to the less hindered side, opposite to the pseudoaxial methyl group, with the result that the natural, but higher-energy, *trans*-fused CD ring junction had been set up. Treatment of the acid (8) with acetic anhydride and sodium acetate gave the enol lactone (9), accompanied by a small amount of the isomer (10).

The remaining skeletal carbon atoms (C-1 to C-4 and C-10) were provided by the Grignard reagent (12), which was prepared[7] from 2-acetylbutyrolactone (11). The Grignard reagent attacked the enol lactone and gave an intermediate diketone (I), which was cyclised without isolation to complete ring B (II). The ketone (II)

(9)   (I)   (II)

had had an opportunity to equilibrate at C-8 in the basic conditions of the reaction; it could therefore be assigned the configuration with the more stable arrangement, namely that with a β-hydrogen. Hydrolysis of the ketal grouping and benzoylation of the C-17 hydroxyl group gave the intermediate (13), which could be converted into a number of different steroids.

The simplest sequence[2] led to the 19-nor steroids. Catalytic hydrogenation of 13 gave the diketone (14), which was cyclised with acid to give the unsaturated ketone (15, R = COPh). In the course of the cyclisation, the equilibrateable position at C-10 had inverted to the more stable arrangement. Hydrolysis of the benzoate group gave 19-nortestosterone (15, R = H).

Another simple transformation[1] was a base-catalysed closure of ring A, to give the dienone (16), followed by a dehydrogenation-hydrogenation rearrangement, to give the aromatic isomer (17, R = COPh). Once again, the preferred *trans* BC ring junction was the one obtained. Alkaline hydrolysis of the benzoate group then gave oestradiol (17, R = H).

A third sequence was used for the introduction of the methyl group (C-19). The saturated ketone was ketalised and the unsaturated ketone alkylated, to give the ketal (18). This reaction is of the same kind as that used by Woodward (p. 60), but the order of the events is reversed. In Woodward's synthesis, the methyl group was already in the molecule (III), and a three-carbon unit was introduced by alkylation of the unsaturated ketone. The unfortunate feature of this reaction, in Woodward's case, was that the major isomer produced (IV) was not the one

(III)   (IV)   (V)

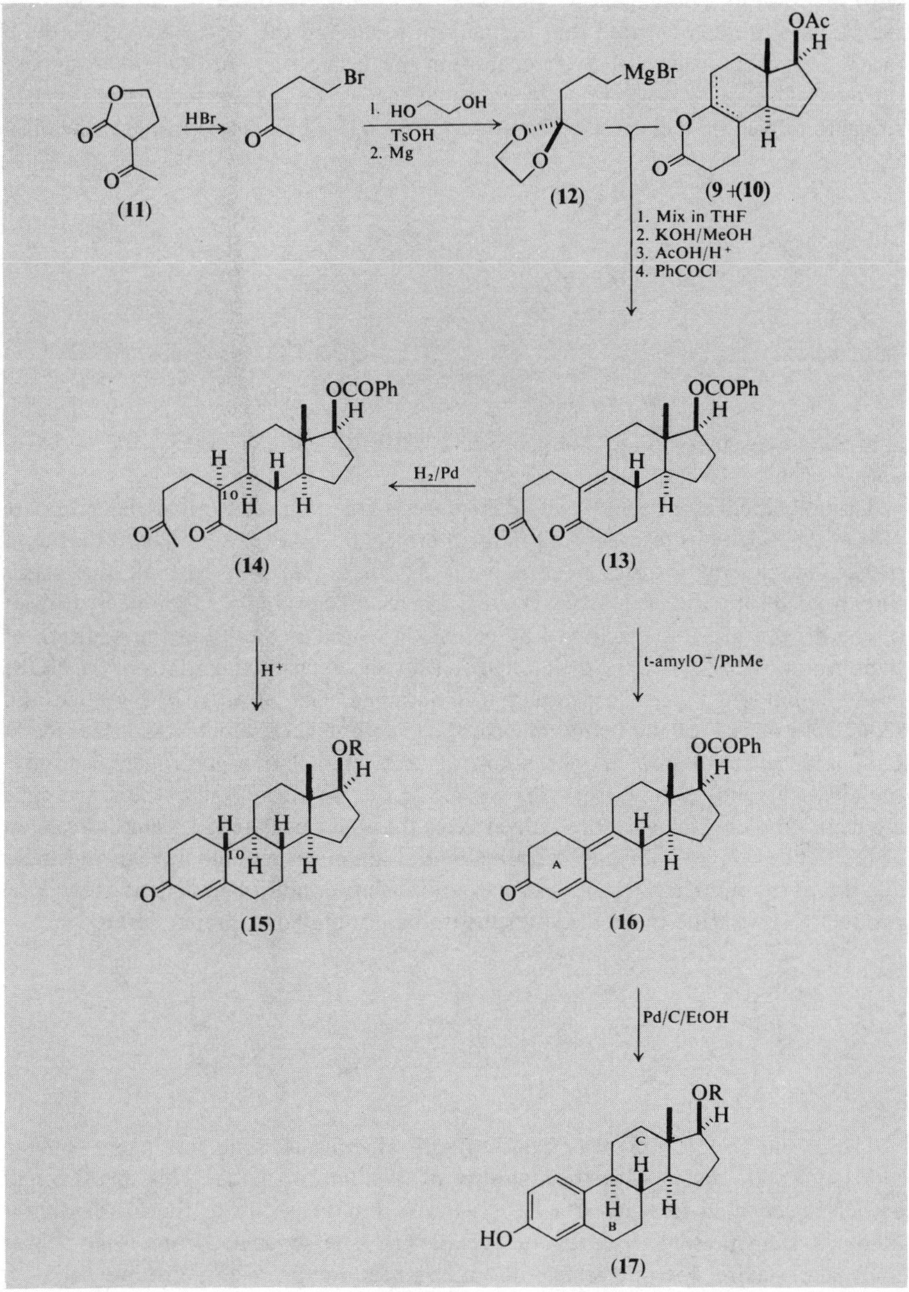
(11)
HBr
Br
1. HO OH
TsOH
2. Mg
MgBr
(12)
OAc
H
(9+(10)
1. Mix in THF
2. KOH/MeOH
3. AcOH/H+
4. PhCOCl
OCOPh
H2/Pd
(14)
(13)
H+
t-amylO−/PhMe
OR
(15)
(16)
A
Pd/C/EtOH
HO
C
B
(17)

with the natural configuration at C-10. By reversing the order of the alkylations, the Velluz synthesis ensured that the major isomer had the methyl group on the β face.[8] This stereochemical preference—for the last group introduced to appear on the β face of the system—is a consequence of the axial attack which causes ring B to move into a chair conformation (VI → VII). This factor in stereochemical

(VI)   (VII)

control was discussed on p. 77 in connection with the Stork synthesis of dehydro-abietic acid.

The alkylated ketone (18) was deketalised and cyclised to give the benzoate (19) of $\Delta^{9,11}$-dehydrotestosterone. This intermediate was very similar to that used in the Woodward synthesis (compound 22 on p. 62) and had all the versatility needed for a wide range of steroid syntheses. For example,[9] alkaline hydrolysis of the benzoate group, followed by oxidation of the alcohol and the addition of hypobromous acid, gave the bromohydrin (20), in which diaxial attack of the HOBr ensured that only a single product was obtained. Oxidation to the ketone and reductive removal of the bromine gave the triketone (21), which was attacked, at C-17 and on the α-side, by the anion of acetylene. Partial hydrogenation gave the allylic alcohol (22). With hydrobromic acid, this alcohol gave the rearranged allylic bromide, which was then converted to the allylic acetate (23). The conversion of the alcohol (22 = VIII) to the bromide (X) took place with allylic rearrangement via the carbonium ion (IX); under equilibrating conditions such as these, the product (X) with the more heavily substituted double bond preponderates.

(VIII)   (IX)   (X)

The allylic acetate (23) was oxidised with an unusual oxidising agent, phenyl iodosodiacetate and a catalytic quantity of osmium tetroxide. This mixture had earlier[10] been used to convert a $\Delta^{17(20)}$-double bond directly to the 17α-hydroxy-20-one system present in cortisone acetate (1). The sequence from 19 to 1 was unusual in having been possible without any need for protecting groups.

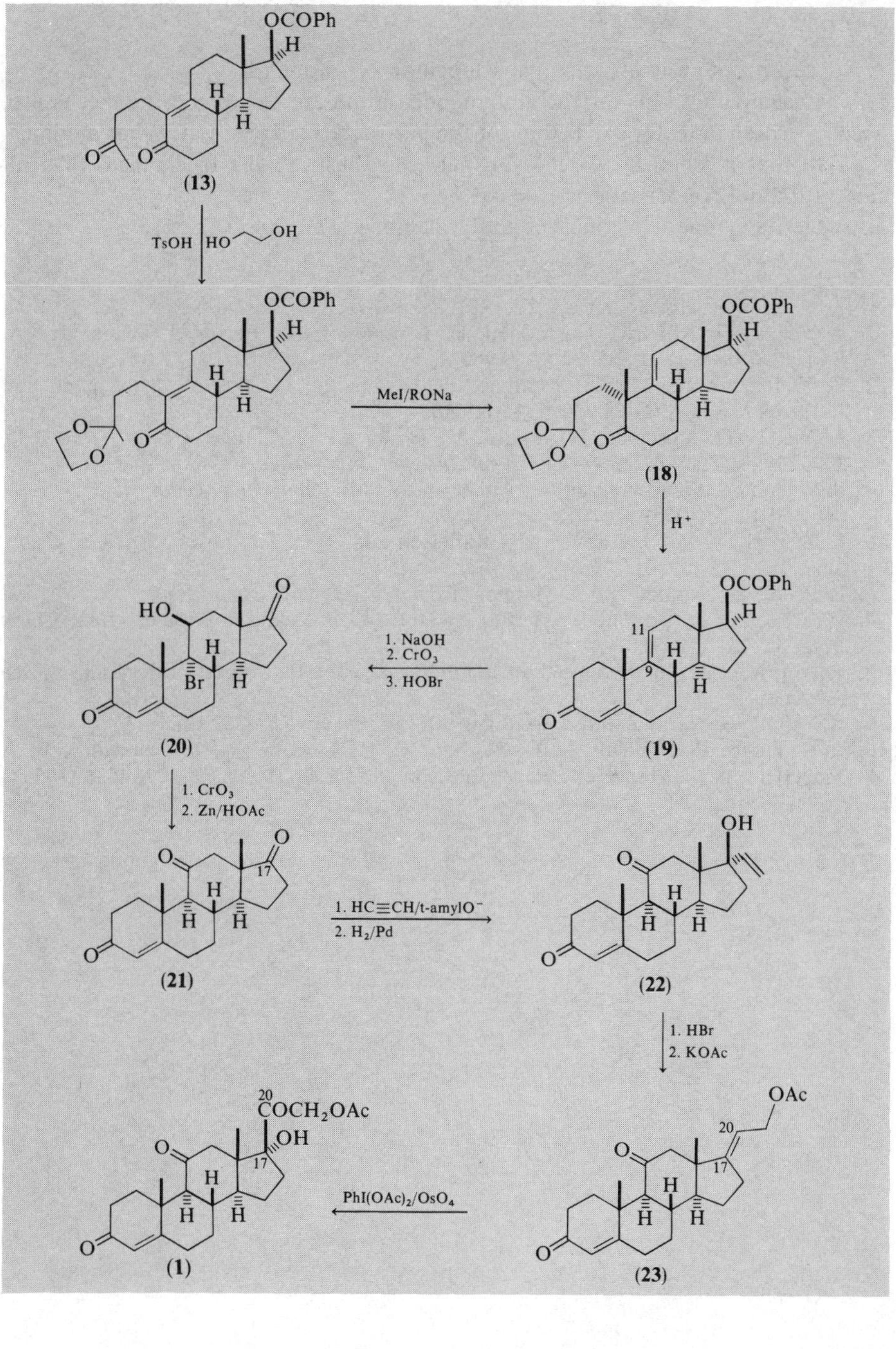
(13)
TsOH
MeI/RONa
(18)
H+
(19)
1. NaOH
2. CrO3
3. HOBr
(20)
1. CrO3
2. Zn/HOAc
(21)
1. HC≡CH/t-amylO⁻
2. H2/Pd
(22)
1. HBr
2. KOAc
(23)
PhI(OAc)2/OsO4
(1)

Questions

1. Suggest a synthesis of 2-methylcyclopentan-1,3-dione (**4**).
2. The usual conditions for the aldol condensation are basic, and the mechanism usually written in textbooks is that for the base-catalysed reaction. What modifications to that mechanism would you make to illustrate the mechanism of acid-catalysed aldol condensations, such as **6** → **7**?
3. Suggest a synthesis of the acetylbutyrolactone (**11**).

References

1. L. Velluz, G. Nominé, J. Mathieu, E. Toromanoff, D. Bertin, J. Tessier and A. Pierdet, *Compt. rend.*, **250**, 1084 (1960).
2. L. Velluz, G. Nominé, J. Mathieu, E. Toromanoff, D. Bertin, M. Vignau and J. Tessier, *Compt. rend.*, **250**, 1511 (1960).
3. L. Velluz, G. Nominé, G. Amiard, V. Torelli and J. Cérède, *Compt. rend.*, **257**, 3086 (1963); L. Velluz, *Angew. Chem. Internat. Edn.*, **4**, 181 (1965).
4. P. Wieland, H. Ueberwasser, G. Anner and K. Miescher, *Helv. Chim. Acta*, **36**, 376, 646 and 1231 (1953).
5. L. B. Barkley, W. S. Knowles, H. Raffelson and Q. E. Thompson, *J. Amer. Chem. Soc.*, **78**, 4111 (1956).
6. P. Bellet, G. Nominé and J. Mathieu, *Compt. rend.*, **263C**, 88 (1966).
7. W. R. Boon, British Patent 558,286 (1943); C. A. Grob and R. Moesch, *Helv. Chim. Acta*, **42**, 728 (1959).
8. This approach was also used in a modification of the Woodward synthesis, see reference 5.
9. L. Velluz, G. Nominé and J. Mathieu, *Angew. Chem.*, **72**, 725 (1960).
10. J. A. Hogg, P. F. Beal, A. H. Nathan, F. H. Lincoln, W. P. Schneider, B. J. Magerlein, A. R. Hanze and R. W. Jackson, *J. Amer. Chem. Soc.*, **77**, 4436 (1955).

Caryophyllene

(1)

The sesquiterpene diene caryophyllene (**1**) presented quite unusual problems in structure determination, largely as a result of the many extensive rearrangements and transannular reactions it undergoes. The synthesis, too, presented particular problems, most notably the need to prepare the two rings, one four membered and the other nine membered; these ring sizes were well known to require special reactions and skills. The synthesis described by Corey[1] is particularly instructive in its solution of the problems posed by these rings. The four-membered ring was prepared by a photochemical coupling. The nine-membered ring, which would have been very difficult to get from an open chain precursor, was set up by opening the central bond of a bicyclic system with six- and five-membered rings fused together, a system much easier to generate.

The first step was the photochemical one and fortunately this proved to be clean and high-yielding, a result not unknown in photochemistry, but the exception rather than the rule. The major product from the photocoupling of isobutylene and cyclohexenone was the *trans*-fused ketone (**2**). This reaction, giving a *trans*-fused ring, has received special attention[2] in its own right, but may well need reinterpretation in the light of the new knowledge of the Woodward-Hoffmann rules[3] and the knowledge that some cyclohexene photochemistry can be explained[4] in terms of the production of *trans*-cyclohexenes.

The *trans*-fused ring system, however, although it was what was eventually wanted in caryophyllene, was much too difficult to preserve when flanked by the enolisable carbonyl group. The compound easiest to isolate was the *cis*-fused isomer (**3**). The methylene group was further activated by condensation with methyl carbonate, and was then methylated. The product (**4**) was a mixture of the two diastereoisomers, which were not separated. The next carbon–carbon bond was inserted, using as the nucleophilic carbon component the acetylenic anion (**5**). The triple bond, having served its function of making the anion (**5**) easy to generate, was removed by hydrogenation, and the acetal was oxidised to give the lactone (6). Dieckmann cyclisation of the ester lactone proceeded smoothly, using as base the methyl sulphinyl carbanion, but did not proceed with more conventional bases such as methoxide. Hydrolysis and decarboxylation gave a crystalline ketone (7); that is, at this stage, one isomer of the mixture of diastereo-isomers had separated. The relative configuration at the starred carbon atoms in this compound was not known, nor, as it happens, did it matter. In any case, since the compound had been subjected to basic conditions, it was likely that equilibration had occurred at both centres as a result of ring opening to the enol (II) of the diketone.

(I)   (II)

Reduction of the ketone (**7**) with Raney nickel gave the two diastereoisomeric alcohols (**8**, R = H), both of which gave monotosylates (**8**, R = Ts), as it is generally easy to prepare primary or secondary tosylates without affecting a tertiary hydroxyl group. One of these two tosylates, on treatment with base, underwent fragmentation and gave the keto-olefin (**9**); the other tosylate gave the *cis* isomer of **9**. The reaction involved is a concerted fragmentation (III, arrows). The stereochemistry of the ring-fusion was not necessarily important, but the relative configuration of the tosyloxy and the neighbouring methyl group was important.

(III)   (IV)   (V)

In the structures III and V the *trans* double bond was being set up from each of the two possible isomers in which the tosyloxy group was *cis* to the methyl group; in IV the *cis* double bond was being set up from one of the possible isomers in which the tosyloxy group was *trans* to the methyl group. In each case there was a reasonably accessible conformation in which coherent overlap would be maintained as the fragmentation proceeded.

This beautifully chosen reaction had now established the ring system of caryophyllene. Base-catalysed equilibration isomerised **9** to **10**, which had the *trans*-fused ring junction, and therefore the natural configuration; it was also, evidently and fortunately, the more stable configuration. A Wittig reaction, again using the methyl sulphinyl carbanion as the base, gave caryophyllene (**1**).

In the course of this synthesis, it was also demonstrated beyond doubt that caryophyllene and isocaryophyllene, an isomer accompanying caryophyllene in natural extracts, differ only in the configuration about the double bond. The *cis* isomer of **9** was taken through the same procedure, which gave isocaryophyllene. When, however, **9** was subjected to the Wittig reaction before epimerisation, another, but this time unnatural, isomer of caryophyllene was produced. The caryophyllenes, then, do not differ in the configuration at the ring fusion.

Questions

1. Examine a model of **6** and suggest a likely course for the Dieckmann reaction on it.
2. Make a model of the other diastereisomer (i.e. not IV) which would lead to the *cis* isomer of **9**, and examine the conformation needed for fragmentation to ensue. Is it likely to be easily attained?

References

1. E. J. Corey, R. B. Mitra and H. Uda, *J. Amer. Chem. Soc.*, **86**, 485 (1964).
2. E. J. Corey, J. D. Bass, R. LeMahieu and R. B. Mitra, *J. Amer. Chem. Soc.*, **86**, 5570 (1964); see also R. Robson, P. W. Grubb and J. A. Barltrop, *J. Chem. Soc.*, 2153 (1964).
3. R. B. Woodward and R. Hoffmann, *Angew. Chem. Internat. Edn.*, **8**, 781 (1969).
4. J. A. Marshall, *Accounts Chem. Res.*, **2**, 33 (1969).

[18] Annulene

(1)

The stability and special properties of benzene and its derivatives were so well known and remarked upon that for many years organic chemistry was divided for pedagogical purposes into two parts—aliphatic chemistry and aromatic chemistry. The advent of mechanistic insights and the extension of the concept of aromaticity have blurred this artificial distinction. One of the outstanding contributions of theory to organic chemistry has been this development of the idea of aromaticity.

The Hückel (4n + 2) rule for aromaticity was first presented in 1931.[1] Cyclobutadiene and cyclooctatetraene had already presented enough difficulty in their preparation to discountenance though not disprove any claim for great stability on their behalf. In the 1950's, however, it became clear that both cyclobutadiene (that is [4]annulene) and cyclooctatetraene ([8]annulene) *might* be unstable simply because of ring strains. (*Annulene* is a general name for the fully conjugated cyclic poly-olefins.) [10]Annulene, [12]annulene, [14]annulene, and [16]annulene could likewise be seen from molecular models to exhibit various strains and steric repulsions which might prevent coplanarity and, hence, reduce orbital overlap. Only at [18]annulene (in. the geometrical isomer (1) with six *trans* double bonds and three *cis* double bonds) were the features of fairly easily attained coplanarity and the possession of a (4n + 2) perimeter both present. The synthesis[2] of this compound revealed that it possessed characteristics common (see below) to other aromatic systems and Hückel's rule thus received its most dramatic support.

The synthesis of [18]annulene presented the problem of making a large ring and, at the same time, keeping to a minimum the number of steps involving poly-olefinic intermediates, which polymerise readily. This was solved by using oxidative coupling of the diacetylene (2), which was readily prepared by the route shown. The coupling of the diacetylene, in this case using the Eglinton procedure, gave a mixture of compounds which was chromatographed on a large column of alumina The first compound to be eluted was the acyclic 'dimer' (9%), 1,5,7,11-dodeca-tetrayne. The next four substances were the cyclic 'trimer' (6%) (3), the cyclic 'tetramer' (6%), the cyclic 'pentamer' (6%), and the cyclic 'hexamer' (2%).

The crystalline product (3) was shown by its ultraviolet spectrum, to be a 1,3-diyne and, by hydrogenation to cyclooctadecane, to have an 18-membered ring. On treatment with base it was isomerised to the fully conjugated tridehydro-[18]annulene (4). Hydrogenation of 4 was stopped with comparative ease so as to give a 31–32% yield of [18]annulene (1).

The structure of the product has been proved by X-ray crystallography. Its n.m.r. spectrum, with 'inside' hydrogens shifted upfield and the 'outside' hydrogens downfield from the normal position of resonance, has led to the diamagnetic ring-current criterion of aromaticity.[3,4] [18]Annulene has been nitrated and acetylated[5]— although ease of substitution is no longer regarded as a *necessary* property of an aromatic compound.

Since this synthesis was completed, many other annulenes and many 4n + 2 and 4n cyclic conjugated ions have been prepared. Most notably, the planar 4n systems have ring currents in the opposite direction: 'inside' hydrogens, if there are any, are at low field and 'outside' hydrogens at high field.[4] Many of the 4n systems show remarkable reactivity; they appear to be positively destabilised. This phenomenon has led to a concept[6] of 'anti-aromaticity', which calls attention to the unfavourable result of introducing cyclic conjugation in such systems. The most elusive and the most sought after of these 4n systems is the subject of the next chapter.

Questions

1. The step **3** → **4**, is reasonably straightforward. Draw out the mechanism.
2. The chemistry of the remaining steps **2** → **3** and **4** → **1** is not straightforward, nor is anything much known about these reactions. Consider, at least, the mechanistic problems which these reactions raise.

References

1. E. Hückel, *Z. Physik.*, **70**, 204 (1931).
2. F. Sondheimer and R. Wolovsky, *J. Amer. Chem. Soc.*, **84**, 260 (1962); F. Sondheimer, R. Wolovsky and Y. Amiel, *ibid.*, 274.
3. L. M. Jackman, F. Sondheimer, Y. Amiel, D. A. Ben-Efraim, Y. Gaoni, R. Wolovsky and A. A. Bothner-By, *J. Amer. Chem. Soc.*, **84**, 4307 (1962).
4. F. Sondheimer, I. C. Calder, J. A. Elix, Y. Gaoni, P. J. Garratt, K. Grohmann, G. di Maio, J. Meyer, M. V. Sargent and R. Wolovsky, Chem. Soc., Special Publication No. 21, *Aromaticity* (1967).
5. I. C. Calder, P. J. Garratt, H. C. Longuet-Higgins, F. Sondheimer and R. Wolovsky, *J. Chem. Soc.* (*C*), 1041 (1967).
6. R. Breslow, J. Brown and J. J. Gajewski, *J. Amer. Chem. Soc.* **89**, 4383 (1967).

Cyclobutadiene

(1)

Academic interest in cyclobutadiene (**1**) and its derivatives is so great that it has inspired, over the last sixty years, an unprecedented amount of work designed to achieve a synthesis of such a simple compound.[1] In the end that goal was achieved so simply, as we shall see, that the immense effort which preceded it looks elephantine. The main problem was the expected instability of the non-aromatic system of **1**: the final step in a synthesis must be a mild reaction if any chance of finding this product was to be entertained.

A product often encountered, when cyclobutadiene might reasonably have been produced, was the dimer (I). It seemed likely that the dimer had been produced

from cyclobutadiene, which had undergone a stereospecific *endo,cis* cyclo-addition to itself, as expected for a Diels-Alder reaction. The dimer (II) was also known and was produced in other reactions in which cyclobutadiene *might* have been an intermediate. It seemed likely that II had been produced when cyclo-butadiene was *not* an intermediate and that, in the reaction shown, for example, it had come from two successive Wurtz reactions which had given the more stable *anti* product (II).[2] This difference in behaviour provided some circumstantial evidence that cyclobutadiene had enjoyed a fleeting existence and indicated that the preferred mode of reaction of free cyclobutadiene is to dimerise* rather than, say, to divide into two acetylenes.

---

* This is only true, of course, if the concentration of cyclobutadiene is not infinitesimal. If the con-centration were so low that collisions very rarely occurred it may well be that other reaction pathways would be preferred.

Such circumstantial evidence was not, of course, convincing proof that cyclobutadiene had indeed been synthesised. The synthesis now to be described was the first to provide more certain evidence. The important reaction was the oxidative release of the hydrocarbon from its complex (**7**); but first we must see how the complex itself was prepared.

Chlorination of cyclooctatetraene (**2**) gave two dichlorides (**3**), which with acetylene dicarboxylic ester gave the pair of adducts (**4**). On heating, **4** gave, by a reverse Diels-Alder reaction, dimethyl phthalate and two more volatile products (**5**) and (**6**), which could be readily separated by fractional distillation.[3]

The *cis*-dichlorocyclobutene (**5**) was dehalogenated by an excess of iron enneacarbonyl, $Fe_2(CO)_9$, and a complex of cyclobutadiene (**7**) was obtained.[4] Complexes of cyclobutadiene were known to be stable. One of the most dramatic demonstrations of the power of molecular orbital theory had been the prediction[5] that such complexes would be relatively stable.

Oxidation of the complex (**7**) released the hydrocarbon. In the absence of a trapping reagent, the dimers (I and II) (in a ratio of 5:1) were obtained. The main product, the dimer (I), was that which was already known from other reactions in which cyclobutadiene might have been produced. However, better evidence that the hydrocarbon enjoyed a fleeting existence came from trapping experiments.[6] When the oxidation was done in the presence of methyl propiolate, the product was the 'Dewar benzene' (**8**). Other dienophiles, such as quinones and fumarate, and several dienes, such as cyclopentadiene, also reacted to give the Diels-Alder adducts expected if cyclobutadiene were present. But the most striking evidence for the existence of *free* cyclobutadiene came from the following experiment.

A solution of the complex (**7**) was added to the oxidising agent, ceric ammonium nitrate, at 0°. The gases evolved were trapped in a liquid nitrogen trap and methyl propiolate was *then* added. On work-up some methyl benzoate was found to be present, indicating that cyclobutadiene, as such, had passed from the reaction flask into the trap.

We can conclude from the results of these experiments that although cyclobutadiene itself can never be isolated (except, just possibly, at very low temperatures), we can nevertheless say that cyclobutadiene has been synthesised.

The annelated derivatives of cyclobutadiene, such as biphenylene, are easily prepared. But the first simple (that is non-annelated) cyclobutadiene to be isolated was the derivative (**13**) containing the vinylogous urethane structure (**14**, arrows). The idea that such overlap would stabilise a cyclobutadiene was due to Roberts.[7] In spite of earlier unsuccessful attempts[8] to prepare cyclobutadienes with such a feature, the successful route[9] was again very short.

The vinylogous urethane (**9**) was brominated to give **10** which, with strong base, gave the ethynylogous urethane (**11**). With boron trifluoride and phenol in ether

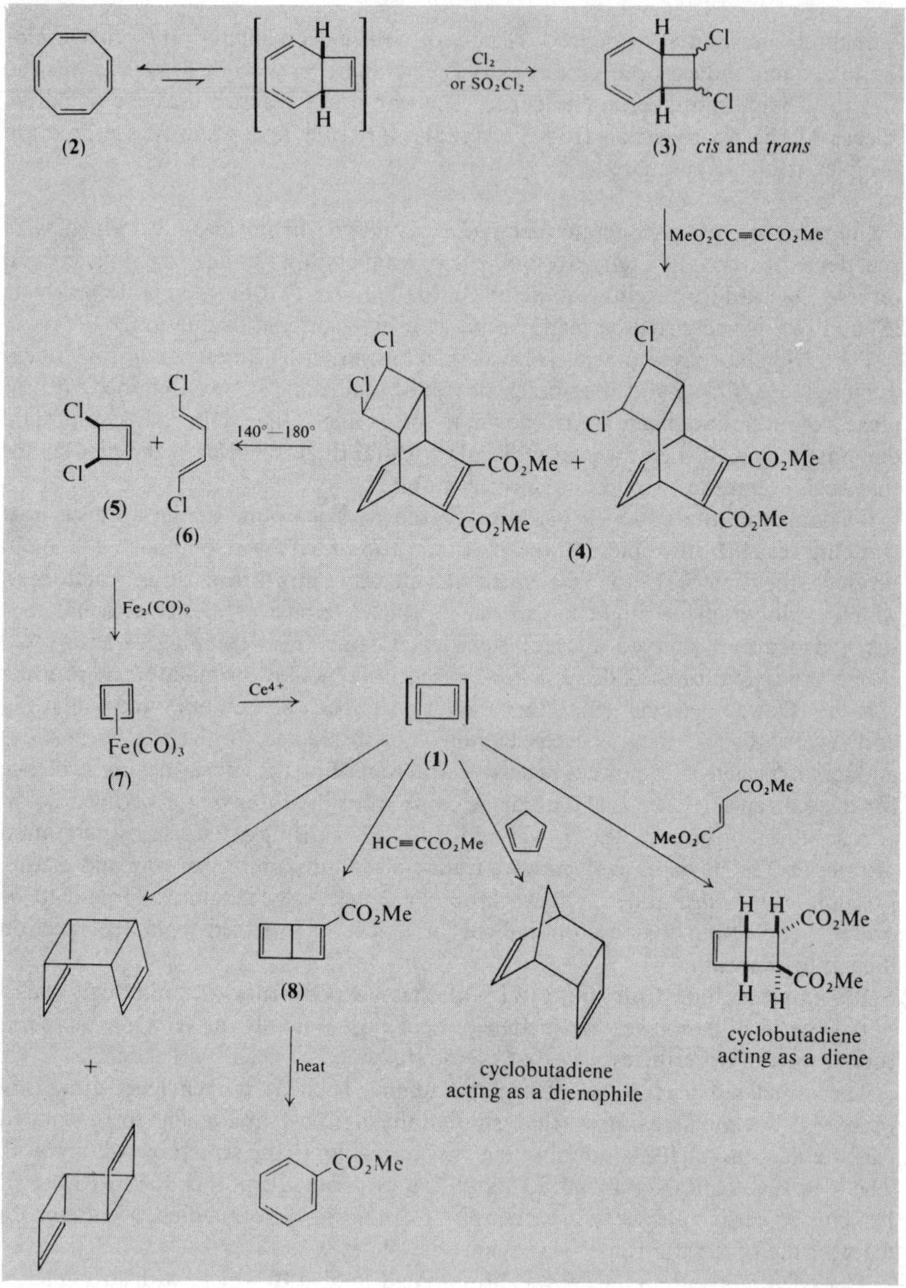

Selected Syntheses

(2)

(3) *cis* and *trans*

$MeO_2CC\equiv CCO_2Me$

(5)

(6)

$140°-180°$

$-CO_2Me$
CO_2Me

(4)

$-CO_2Me$
CO_2Me

$Fe_2(CO)_9$

$Fe(CO)_3$
(7)

Ce^{4+}

(1)

$HC\equiv CCO_2Me$

CO_2Me
(8)

heat

CO_2Me

cyclobutadiene
acting as a dienophile

MeO_2C
CO_2Me

CO_2Me
CO_2Me

cyclobutadiene
acting as a diene

the ynamine (11) 'dimerised' to give the salt (12). This reaction probably involves the concerted cycloaddition (III + IV, solid arrows) of the 'vinyl cation' (III) to the ynamine (IV), aided by the polarisability (III + IV, dotted arrows). This

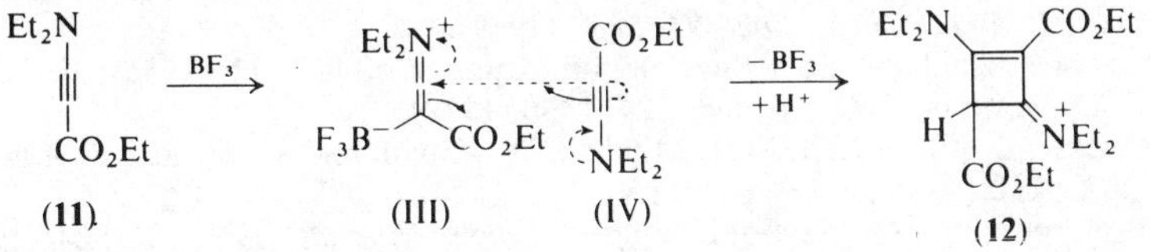

(11) (III) (IV) (12)

cycloaddition is 'allowed' because the ketene-like arrangement of orbitals in III makes possible *antarafacial* attack on III. The reaction is, then, of the $(\pi 2s + \pi 2a)$ type, which is thermally allowed to be concerted.[10]

When 12 was treated with sodium hydride in an aprotic solvent, a yellow colour, due to 13, was observed. Evaporation of the solvent gave a residue from which the cyclobutadiene (13) was isolated by crystallisation.

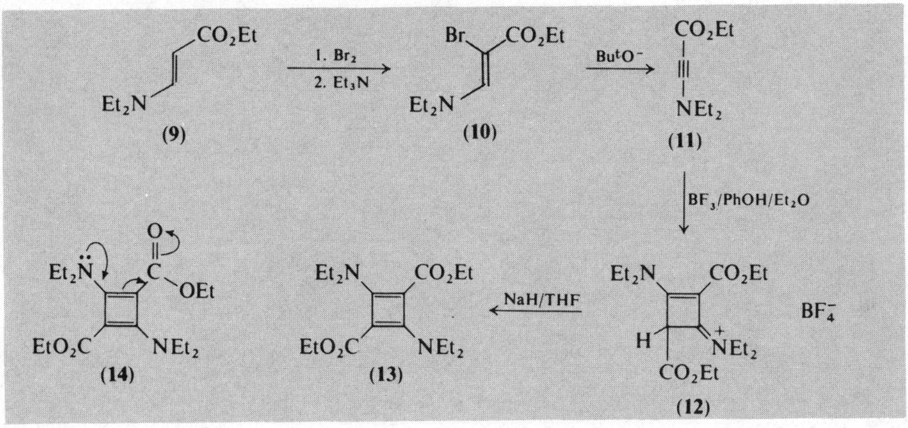

(9) (10) (11)

(14) (13) (12)

Question

Account for the formation of *acyclic* 6, but *cyclic* 5, when the mixture (4) was heated.

References

1. M. P. Cava and M. J. Mitchell, *Cyclobutadiene and Related Compounds*, Academic Press, New York, 1967; R. Criegee, *Bull. Soc. chim. France*, 1 (1965).
2. M. Avram, I. G. Dinulescu, E. Marica, G. Mateescu, E. Sliam and C. D. Nenitzescu, *Chem. Ber.*, 97, 382 (1964).
3. M. Avram, I. Dinulescu, M. Elian, M. Farcasiu, E. Marica, G. Mateescu and C. D. Nenitzescu, *Chem. Ber.*, 97, 372 (1964).

4. G. F. Emerson, L. Watts and R. Pettit, *J. Amer. Chem. Soc.*, **87**, 131 (1965).
5. H. C. Longuet-Higgins and L. E. Orgel, *J. Chem. Soc.*, 1969 (1956).
6. L. Watts, J. D. Fitzpatrick and R. Pettit, *J. Amer. Chem. Soc.*, **87**, 3253 (1965); L. Watts, J. D. Fitzpatrick and R. Pettit, *J. Amer. Chem. Soc.*, **88**, 623 (1966); J. C. Barborak, L. Watts and R. Pettit, *ibid.*, 1328; see also P. Reeves, J. Henery and R. Pettit, *J. Amer. Chem. Soc.*, **91**, 5888 (1969).
7. J. D. Roberts, Chem. Soc. Special Publication No. 12, 111 (1958); S. L. Manatt and J. D. Roberts, *J. Org. Chem.*, **24**, 1336 (1959).
8. R. Breslow, D. Kivelevich, M. J. Mitchell, W. Fabian and K. Wendel, *J. Amer. Chem. Soc.*, **87**, 5132 (1965).
9. R. Gompper and G. Seybold, *Angew. Chem. Internat. Edn.*, **7**, 824 (1968). M. Neuenschwander and A. Niederhauser, *Chimia*, **22**, 491 (1968), *Helv. chim. Acta*, **53**, 519 (1970).
10. R. B. Woodward and R. Hoffmann, *Angew Chem. Internat. Edn.*, **8**, 781 (1969).

Adamantanes

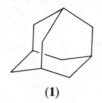

(1)

Adamantane **(1)** is a compound found in small quantities in petroleum and was first synthesised by Prelog.[1] Although the original routes have been improved, adamantane itself is best prepared by a method which was discovered accidentally by Schleyer. He had been studying the acid-catalysed rearrangement of tetra-hydrocyclopentadienedimer **(2)**, a readily available hydrocarbon. In sulphuric

$$\xrightarrow{\text{H}_2\text{SO}_4}$$

(2)　　　　　　　　　　　　　　　　　　**(I)**

acid (99·9%) this gave the *exo* isomer (I), but with aluminium chloride a small amount of adamantane **(1)** was obtained in addition to I. The best yields (19%) were obtained by stirring **2** with a large excess of aluminium bromide, together with a little s-butyl bromide and hydrobromic acid, at room temperature for two days.[2]

Br·

$$\xrightarrow[\text{HBr}]{\text{AlBr}_3}$$
$$\xleftarrow{\text{AlCl}_3}$$

(2)　　　　　　　　**(1)**　　　　　　　　**(3)**

The mechanism of this extensive transformation **(2 → 1)** is, of course, too complicated for us to be sure of what happens. The s-butyl bromide with aluminium bromide will give rise to a carbonium ion. One of the less emphasised capacities of carbonium ions is their capacity to abstract hydride:

$$-\underset{|}{\overset{|}{\text{C}}}{}^{+} \quad H\!-\!\underset{|}{\overset{|}{\text{C}}}\!- \quad \longrightarrow \quad -\underset{|}{\overset{|}{\text{C}}}\!-\!H \quad {}^{+}\underset{|}{\overset{|}{\text{C}}}\!-$$

This process is very important industrially and takes place, for example, in the 'reforming' process whereby straight chain alkanes are isomerised to branched chain alkanes.

In the case of the hydrocarbon (2), hydride abstraction, initiated by the s-butyl cation, can take place from several sites, and a large number of carbonium ion rearrangements might ensue.

One possible route, and the shortest involving only reasonable shifts of hydride and alkyl groups, and hydrogen transfers, is as follows:

This particular route was worked out[3] with the help of a computer, in the course of a study in which the much simpler transformation of 'twistane' (**3**), another $C_{10}H_{16}$ hydrocarbon, to adamantane was examined. With aluminium chloride this hydrocarbon gave adamantane in good yield (85 %), not perhaps surprisingly, in view of the much shorter route from **3** to **1**.

The principle involved in this synthesis is unique—that given time and the right conditions, a mixture of compounds will, at least partly, sink into a trough in the energy surface, in spite of the fact that several steps involve an unfavourable rearrangement from tertiary carbonium ions to secondary carbonium ions. Indeed it has been found[4] that almost any hydrocarbon will give adamantanes, by a suitable number of fragmentations, alkylations and rearrangements, on treatment with aluminium halides. The trough is presumably deep because of the absence of strain in the rings: adamantane is a piece taken out of the diamond lattice.[5]

It is easy enough to extend this concept to larger pieces of the diamond lattice and to aim at the synthesis of such compounds as **4** and **5**, which are sometimes

$C_{14}H_{20}$ AlCl₃ →   (**4**) $C_{14}H_{20}$

Me   Me   AlBr₃ → $C_{18}H_{24}$ (**5**) $C_{18}H_{24}$

called adamantologues of adamantane. Schleyer has actually synthesised these compounds by preparing an isomeric hydrocarbon and then rearranging it under conditions similar to those used for adamantane itself.[6]

As it happens the next higher adamantologue was not obtained when a $C_{22}H_{28}$ hydrocarbon was subjected to rearrangement, but a 'similar to, but inferior to . . .

non-standard' adamantane, which departed slightly from the diamond lattice, was found instead.[7]

This kind of side-track obviously makes difficult the future extensions of this synthetic method. The problem is most easily appreciated if one considers that the isolation of crystalline, unfunctionalised hydrocarbons, such as **4** and **5**, is only the beginning of the work involved in showing that adamantanes have been formed.

Questions

1. Having studied the sequence on p. 150 for the isomerisation of **2** to **1**, consider what the 'shorter route' from **3** to **1** is, and identify where it enters the sequence illustrated.
2. How would you synthesise the precursor of **4**?
3. For those who actually *like* nomenclature: the $C_{22}H_{28}$ hydrocarbon 'similar to', but not actually, an adamantane, is called nonacyclo[11,7,1,1^{2,18},0^{3,16},0^{4,13}, 0^{5,10},0^{6,14},0^{7,11},0^{15,20}]docosane. Draw the structure in a realistic looking way.

References

1. V. Prelog and R. Seiwerth, *Ber.*, **74**, 1644 (1941).
2. P. von R. Schleyer and M. M. Donaldson, *J. Amer. Chem. Soc.*, **82**, 4645 (1960); for a new catalyst system, giving nearly quantitative yields of adamantane, see D. E. Johnston, M. A. McKervey and J. J. Rooney, *J. Amer. Chem. Soc.*, **93**, 2798 (1971).
3. H. W. Whitlock and M. W. Siefken, *J. Amer. Chem. Soc.*, **90**, 4929 (1968).
4. M. Nomura, P. von R. Schleyer and A. A. Arz, *J. Amer. Chem. Soc.*, **89**, 3657 (1967).
5. But see, for some discussion of the concept of strain in admantanes, P. von R. Schleyer, J. E. Williams and K. R. Blanchard, *J. Amer. Chem. Soc.*, **92**, 2377 (1970), and M. Månsson, N. Rapport and E. F. Westrum, *ibid.*, p. 7296.
6. C. Cupas, P. von R. Schleyer and D. J. Trecker, *J. Amer. Chem. Soc.*, **87**, 917 (1965); V. Z. Williams, P. von R. Schleyer, G. J. Gleicher and L. B. Rodewald, *ibid.*, **88**, 3862 (1966).
7. P. von R. Schleyer, E. Osawa and M. G. B. Drew, *J. Amer. Chem. Soc.*, **90**, 5034 (1968).

Cyclopropanones

(1)

Cyclopropanones have appeared in the literature of organic chemistry for many years, but only recently have syntheses been done which could unambiguously be demonstrated to have given isolable cyclopropanones. The reason for the interest in cyclopropanones stems largely from the likelihood that they are intermediates in the Favorskii reaction;[1] it is necessary, in order to show that such intermediates are involved, to show also that they would give the products required of them.

Two highly successful routes to cyclopropanones were developed by Turro and Hammond.[2] Tetramethylcyclopropanone **(3)** was produced by photolysis of tetramethylcyclobutanedione **(2)**. The unchanged starting material could be

removed, since it crystallised on cooling. The product **(3)** could then be co-distilled (10°/20 Torr.) with the solvent, along with small amounts of the products of unavoidable over-irradiation. The loss of carbon monoxide from cyclic ketones is a general reaction and in this case was probably assisted by the formation of a well-delocalised intermediate, possibly the diradical.

The unsubstituted cyclopropanone **(1)** was prepared by mixing a methylene dichloride solution of diazomethane with an excess of ketene at $-130°$. Nitrogen was rapidly evolved. The excess ketene could be removed at $-78°$ under vacuum and the residue flash-distilled and kept at $-78°$. In this way, clean solutions of cyclopropanone **(1)** in methylene dichloride were prepared. The reaction probably involved nucleophilic attack by diazomethane on ketene, followed by cyclisation, in a process resembling the Favorskii reaction. The cyclopropanone was readily identified by its infrared spectrum (ν_{max} 1813 cm^{-1}) and its n.m.r. spectrum (a singlet at τ 8·35).

Cyclopropanones are very reactive molecules, particularly towards nucleophiles: with methanol, hemiketals were produced quantitatively and rapidly, and with diazomethane, ring expansion was readily achieved:

With strong base, dimethylcyclopropanone *did* open[3] to give the expected 'Favorskii' product:

$$\longrightarrow Me_3CCO_2Me$$

There is now considerable evidence that the cyclic structure was correct, but it seems that there is a low-activation-energy electrocyclic conversion to an open-chain zwitterionic structure (for example 3 ⇌ 1). One striking piece of evidence for this isomerism was the trapping of the zwitterion with dienes. Thus, 3 with furan (II) gave the adduct (III), presumably via the symmetry-allowed cycloaddition of an allyl cation to a diene.

The instability of the cyclopropanone ring could be controlled by hindering the attack of any nucleophile. This was done in a synthesis of *trans*-2,3-di-t-butyl-cyclopropanone,[4] again using a reaction of the Favorskii type.

The hindered base used was not able to penetrate to the carbonyl group of 4, which was protected by the large t-butyl groups; further reaction, analogous to the attack of methanol on **1**, was therefore not observed. The product (**4**) was crystalline and was thermally quite stable in the absence of small nucleophiles.[4] It was subsequently resolved,[6] and the optically active cyclopropanone was shown to undergo racemisation (on heating to 80°) by a mechanism of ring-opening and closing, of the kind (**3** ⇌ **1**) mentioned above:

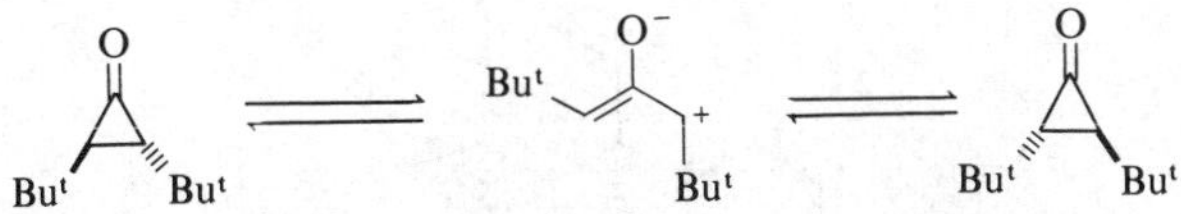

The syntheses just described complete a section on molecules which have been synthesised because of some specially interesting feature which they possess. The last syntheses in this book are some particularly distinguished examples from natural product chemistry.

References

1. A. S. Kende, *Organic Reactions*, **11**, 261, Wiley, New York, 1960.
2. N. J. Turro and W. B. Hammond, *Tetrahedron*, **24**, 6017 (1968).
3. W. B. Hammond and N. J. Turro, *J. Amer. Chem. Soc.*, **88**, 2880 (1966).
4. J. F. Pazos and F. D. Greene, *J. Amer. Chem. Soc.*, **89**, 1030 (1967).
5. For a review of cyclopropanone chemistry see N. J. Turro, *Accounts Chem. Res.*, **2**, 25 (1969).
6. D. B. Sclove, J. F. Pazos, R. L. Camp and F. D. Greene, *J. Amer. Chem. Soc.*, **92**, 7488 (1970)

A Third Steroid Synthesis

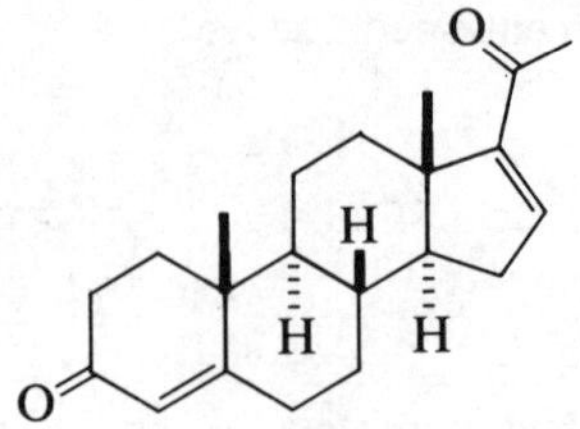

(1)

W. S. Johnson has been responsible for a number of steroid syntheses, but the most remarkable of these came as a result of an extended study[1] of polyolefin cyclisations. His work started from the premise, first voiced in 1955 by both Stork and Eschenmoser, that the preferred mode of cyclisation of squalene, both in nature and in the laboratory, would involve chair-like transition states. It is now known that the actual cyclisation in nature is initiated by the opening of the oxide ring of squalene oxide (I):

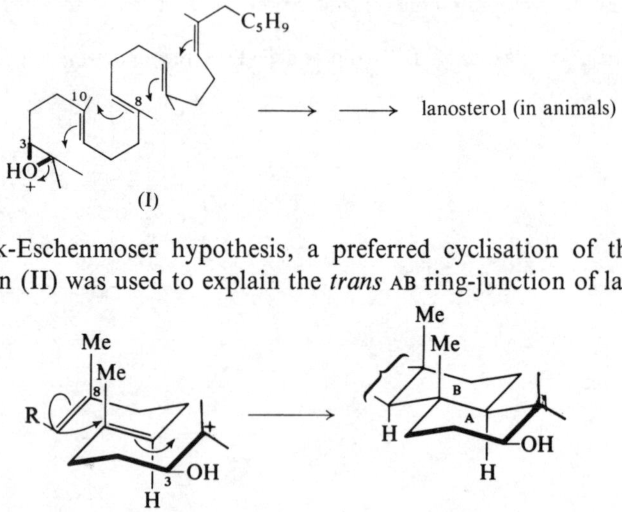

lanosterol (in animals)

(I)

In the Stork-Eschenmoser hypothesis, a preferred cyclisation of the chair-like conformation (II) was used to explain the *trans* AB ring-junction of lanosterol.

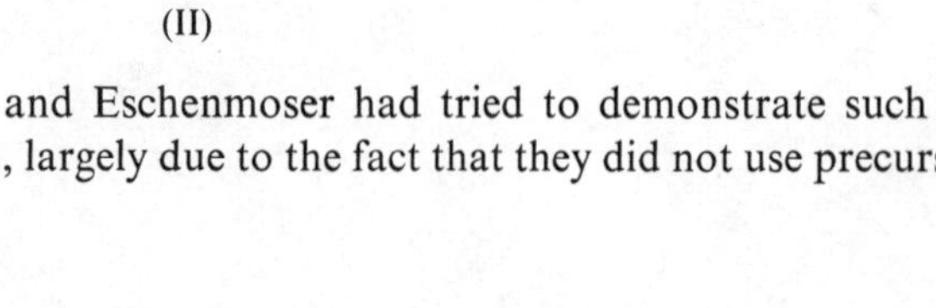

(II)

Both Stork and Eschenmoser had tried to demonstrate such a process in the laboratory but, largely due to the fact that they did not use precursors which would generate specific carbonium ions (such as II), they were unable to show the expected preference for stereospecific *trans* addition to the central double bond of compounds like II. Johnson's achievement was, first, to find the conditions for such stereospecific reactions and, second, to develop these into workable syntheses. Similar considerations also led van Tamelen to develop biosynthetically patterned syntheses, most notably of the onocerins and hopenone.[2]

In the first stage of his work, Johnson found that the dienes (III) and (IV) cyclised in formic acid to give *trans* and *cis* decalin derivatives, such as the formates (V) and (VI). Most importantly he found that III gave no *cis* decalin derivatives and IV gave no *trans* decalin derivatives.

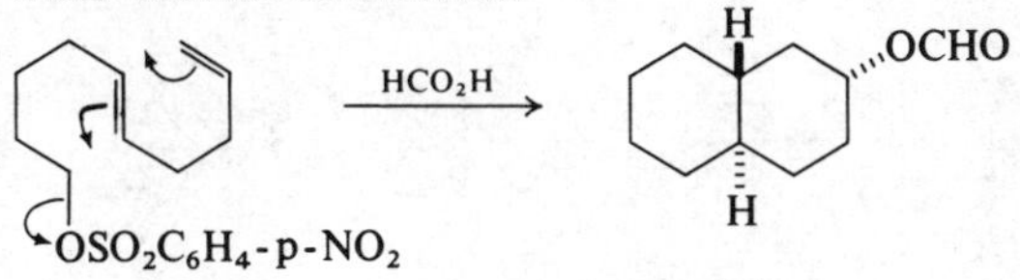

(III)   (V)

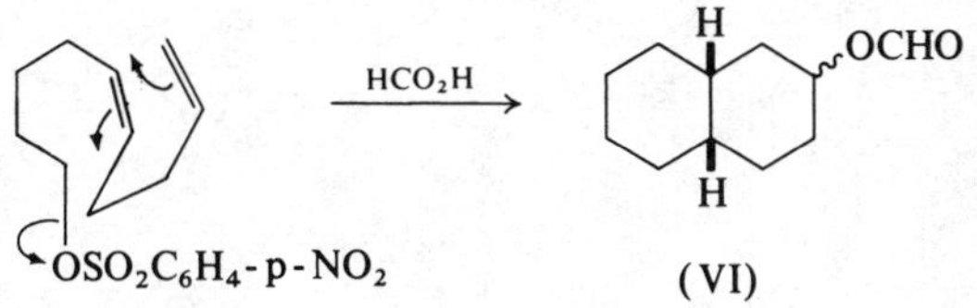

(IV)   (VI)

He then proceeded to examine a great many systems in an effort to find compounds and conditions which would extend the number of stereospecific steps. The most dramatic of these was the cyclisation of the ketal (VII) with stannic chloride to give the tetracyclic product (VIII) in 30 per cent yield.[3] The product had four ring junctions, each of which is specifically *trans*, and six of its seven chiral centres were set up stereospecifically in one step.

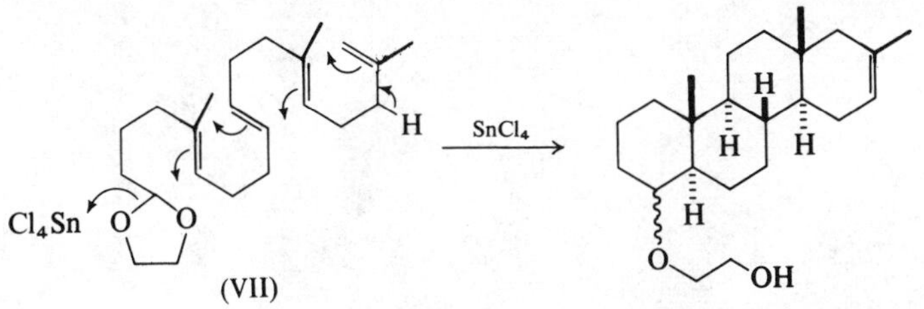

(VII)   (VIII)

Another only slightly less dramatic example, however, setting up five centres in one step, had the advantage that the product could readily be transformed into steroidal materials, and it is this synthesis[4] which will be described below. In a

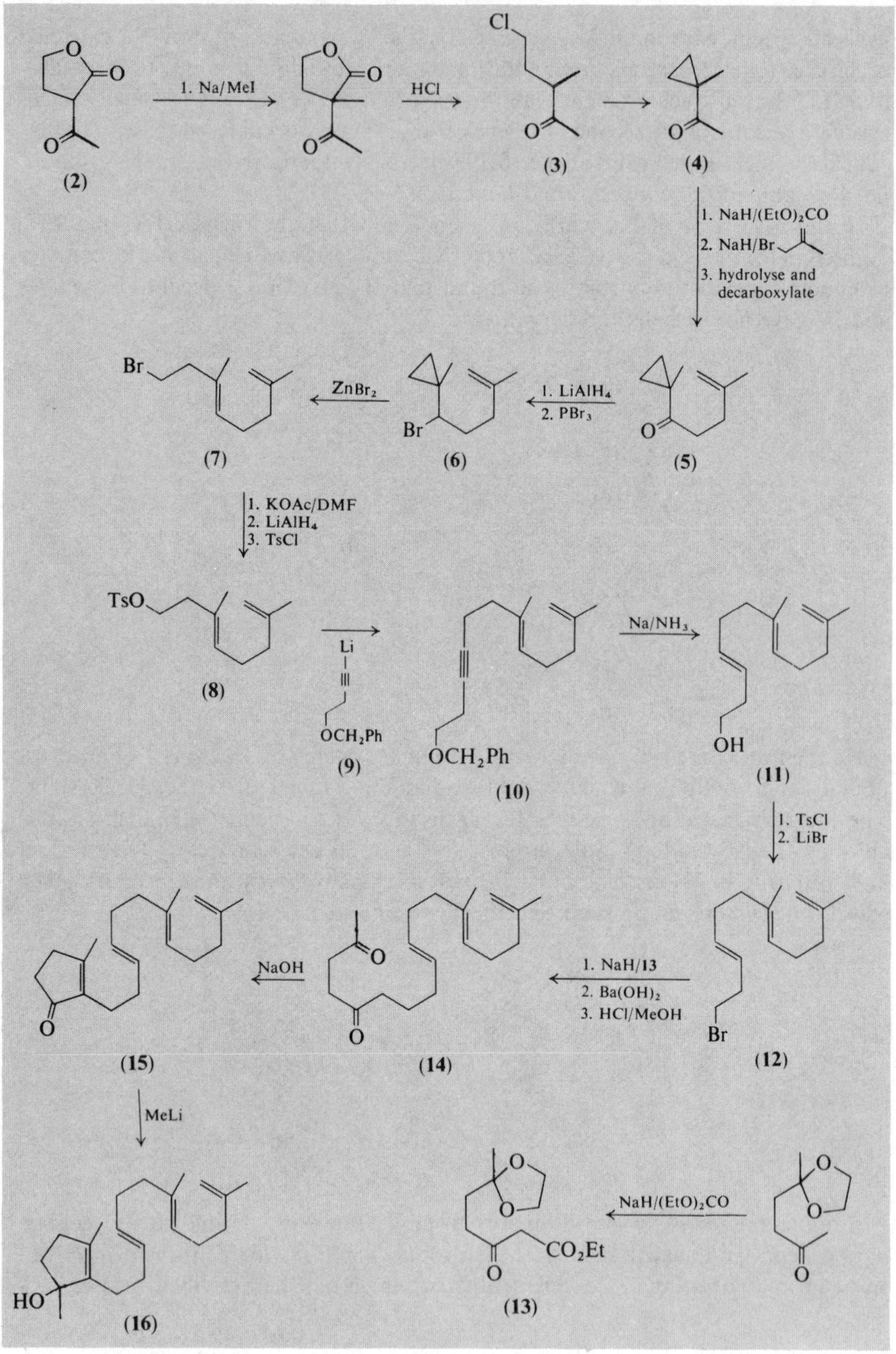

1. Na/MeI
HCl
(2)
(3)
(4)
1. NaH/(EtO)₂CO
2. NaH/Br
3. hydrolyse and decarboxylate
1. LiAlH₄
2. PBr₃
ZnBr₂
(5)
(6)
(7)
1. KOAc/DMF
2. LiAlH₄
3. TsCl
TsO
(8)
OCH₂Ph
(9)
(10)
Na/NH₃
OH
(11)
1. TsCl
2. LiBr
Br
(12)
1. NaH/13
2. Ba(OH)₂
3. HCl/MeOH
(14)
NaOH
(15)
MeLi
HO
(16)
NaH/(EtO)₂CO
CO₂Et
(13)

cyclisation step, such as that of III → V, it was the configuration about the double bond which would determine the configuration at the ring junction. This was a consequence of both the concertedness of the *trans* 'addition' to the double bonds and the chair-like conformation adopted by preference in the transition state. For synthesis, then, at least one requirement was a polyolefin containing only the isomer with the desired configuration about the double bonds. The first stages of the synthesis were concerned with the preparation of such a polyolefin (namely **16**) and involved the development of a new process (**6** → **7**) which would set up one of the double bonds in a stereoselective manner.

The β-ketolactone (**2**) was alkylated and the product, with hydrogen chloride, was converted to the chloride (**3**), which, with base, gave the cyclopropyl ketone (**4**). This ketone was activated by condensation with diethyl carbonate, alkylated with 2-methylallyl bromide, and decarboxylated. Reduction of the ketone (**5**) gave the alcohol, which was converted to the corresponding bromide (**6**). With only one chiral centre, no problems due to mixtures of diastereoisomers had yet arisen. After much experimentation, conditions were found (zinc bromide in ether at 0°) which gave the *trans* bromo-olefin (**7**) containing only a trace of the *cis* isomer.[6]

This high stereoselectivity was a result of the constraints present in the transition state for the process (IX). It is likely that the carbon–carbon bond and the carbon–bromine bond in IX were broken in a concerted fashion and probably, therefore,

(IX)

in a *trans* and co-planar arrangement. There were two conformations possible for the transition state within these constraints: X and XI. Plainly X was the

(X)   (XI)

preferred conformation, since the cyclopropyl group eclipses only a hydrogen atom. It was also X which would lead to the *trans* configuration about the double bond, the configuration we observe in the product (**7**).

The bromide (7) was converted to the alcohol and that to its *p*-toluenesulphonate (8). The sulphonate was, in turn, displaced with the anion of the acetylene (9) to create a new carbon–carbon bond in 10. Sodium in liquid ammonia removed the protecting group and partially reduced the triple bond stereoselectively to the *trans*-trienol (11). This alcohol was converted to its bromide (12), which was used to alkylate the β-ketoester (13). Hydrolysis and decarboxylation gave 14, which was cyclised to give the ketone (15). Treatment with methyl lithium gave the tertiary alcohol (16).

With trifluoroacetic acid (a strong but not very nucleophilic acid) in methylene dichloride, this alcohol underwent cyclisation to give, in 30 per cent yield, the crystalline diene (17). When the crude triene (17) was treated first with osmium tetroxide, the resultant tetraol cleaved with lead tetra-acetate, and the triketo-aldehyde cyclised with alkali, racemic 16,17-dehydroprogesterone (1) was obtained.

That so much of 17 was produced in the cyclisation step implies a high degree of organisation in the transition state, since *trans* addition at both of the double bonds has taken place. No doubt the conformation (XII) is frequently achieved and cyclisation from it is then easy. Nevertheless, the participation of so remote a function as the terminal double bond in the bonding of C-9 to C-10 is remarkable. The evidence that such cyclisations are stereospecific—and stereospecific in the right sense—has added much weight to the original hypothesis that the enzyme system which converts squalene to lanosterol is helping what is, in any case, a favourable process.

Having demonstrated that suitable polyolefins can be made to cyclise in this way and that steroidal products can be prepared, Johnson has gone on to develop[7] an improved synthesis of the starting material (16), and has also found even better conditions for the cyclisation step.

The newer route began with methallyl alcohol (18), the vinyl ether of which underwent a Claisen rearrangement:

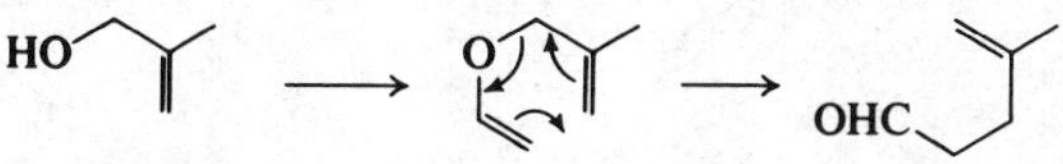

(18)　　　　　　　　　　　　　　　　(19)

to give the aldehyde (19). Isopropenyl lithium reacted with this aldehyde to give the dienol (20). When this dienol was treated with thionyl chloride, it gave, stereoselectively, the *trans* diene chloride (21). One explanation for the stereoselectivity is that the transition state for the intramolecular reaction adopts a chair-like shape (XIII or XIV) in which the substituent adopts an equatorial position (XIII) when it leads to a *trans* product, but would have to adopt an axial position (XIV) if it led to the *cis* product. The former was thus the preferred pathway.

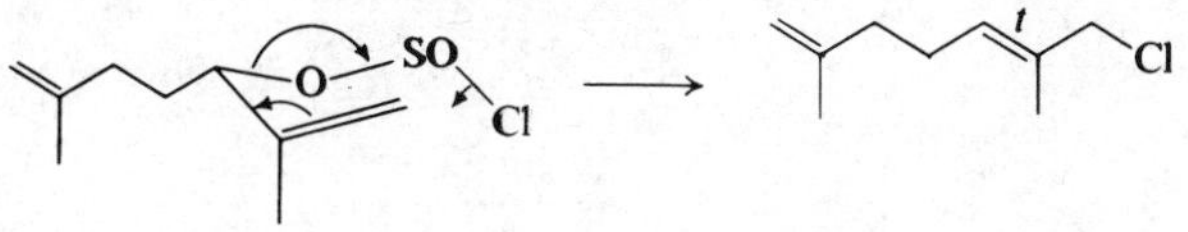

(XIII)

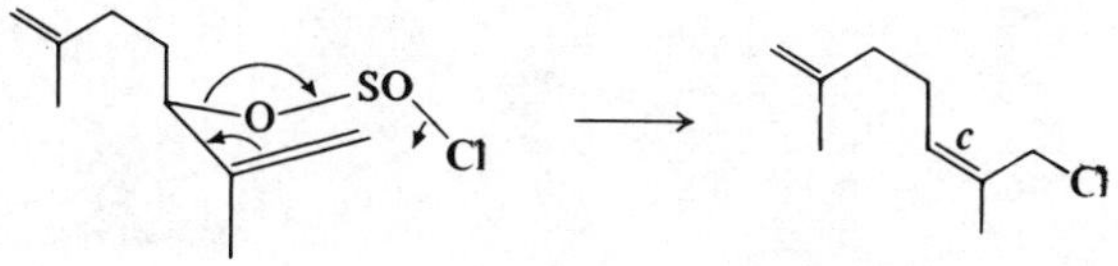

(XIV)

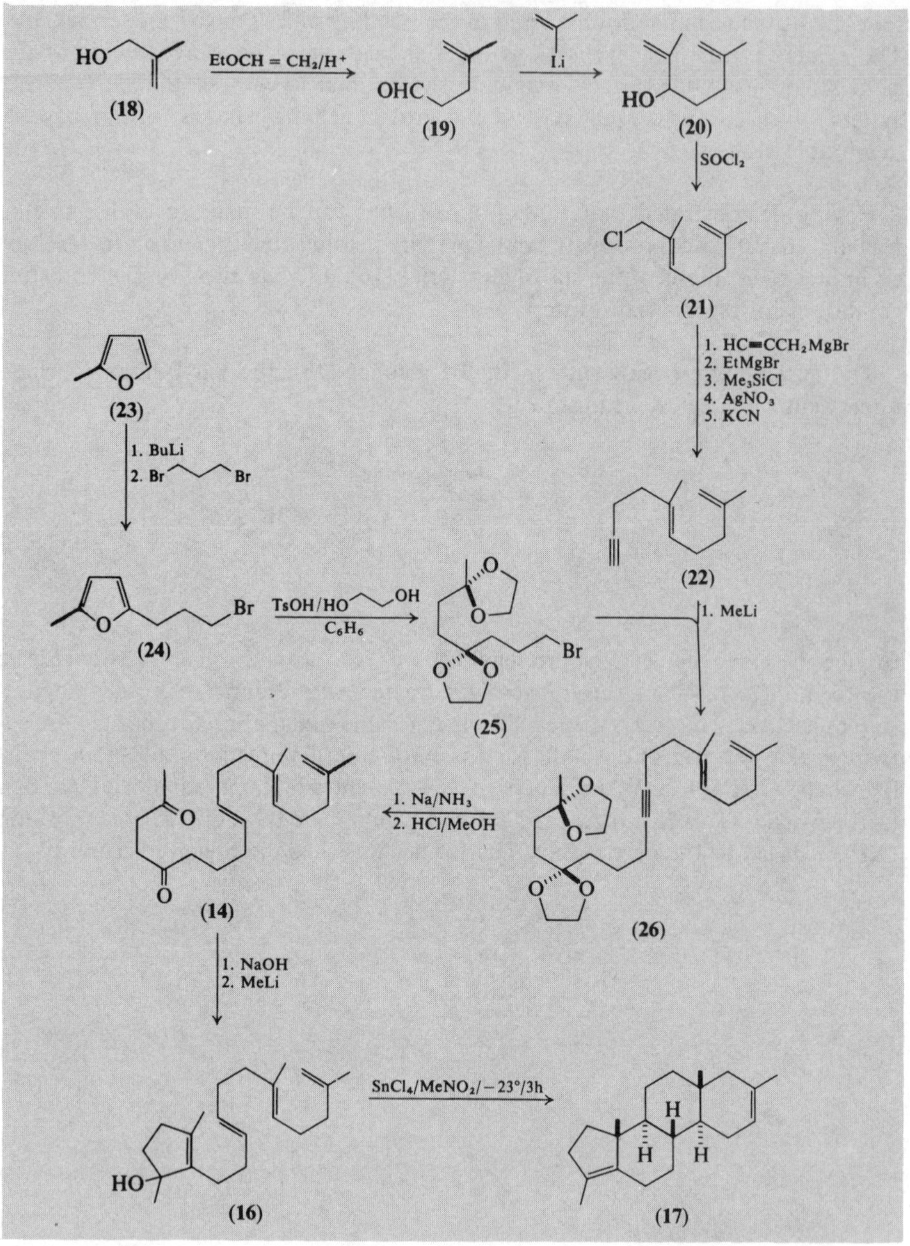
HO
(18)
EtOCH = CH₂/H⁺
OHC
(19)
Li
HO
(20)
SOCl₂
Cl
(21)
1. HC≡CCH₂MgBr
2. EtMgBr
3. Me₃SiCl
4. AgNO₃
5. KCN
(22)
(23)
1. BuLi
2. Br Br
Br
(24)
TsOH/HO OH
C₆H₆
Br
(25)
1. MeLi
(26)
1. Na/NH₃
2. HCl/MeOH
(14)
1. NaOH
2. MeLi
HO
(16)
SnCl₄/MeNO₂/−23°/3h
H
H
H
(17)

The allylic chloride in **21** was displaced with the propargyl Grignard reagent.[8] (Because this nucleophilic carbon reagent is an ambident nucleophile, a mixture of acetylenic and allenic products is obtained in this kind of reaction, with the former predominating;

only the acetylenic product can be silylated, and in this condition separated from the small amount of allene product.) Removal of the silyl groups was carried out, using Arens' procedure (see p. 168), to give the dieneyne (**22**).

The other half of the molecule was built up from 2-methylfuran (**23**). n-Butyl lithium gave an intermediate organometallic species (presumably XV but possibly XVI), which was alkylated with an excess of 1,3-dibromopropane. Although

(XV) (XVI)

furans are masked 1,4-diketones, and can be prepared from 1,4-diketones, the hydrolysis of a furan to a 1,4-diketone is not usually a clean reaction. In this case, by including ethylene glycol in the reaction mixture, the furan (**24**) was converted in good yield to the bisketal (**25**) of the 1,4-diketone.

The two halves of the molecule were then ready to be joined. The anion of the acetylene (**22**) was used to displace the bromine of **25**; sodium and ammonia reduction of the triple bond gave the bisketal (**26**); and acid hydrolysis removed the ketal groups to give the same diketone (**14**) as that obtained in the earlier route.

The improvement in the cyclisation step (**16** → **17**) took the form of a change of solvent, to nitromethane, and change of acid catalyst, to stannic chloride. The yield of the hydrocarbon (**17**) was then 70 per cent. The yields obtained in the new route, although not yet optimum, indicated that 100 g of the starting material (**18**) would give about 10 g of the 16,17-dehydroprogesterone (**1**). This synthesis is an improvement partly because the reactions used gave better yields than those in the earlier route, and also because the new route is convergent.

Questions

1. In the synthesis of VII[3] the alcohol **11** was converted in several steps to the chloroketone XIII, which was treated at low temperature with the Grignard reagent (XIV), to give a single diastereosiomer (XV).

The resultant chlorohydrin (XV), with alkali, gave a single epoxide, which re-opened with hydrobromic acid to give a bromohydrin in which, presumably, inversion of configuration at both C-6 and C-10 had taken place. Treatment of the bromohydrin with zinc powder gave the olefin, which was predominantly the desired one with a *trans* double bond. This type of stereospecific olefin synthesis is due to Cornforth and was used by him in the synthesis of squalene.[9]

The Grignard reaction and the zinc-promoted elimination are obviously stereoselective. Which diastereoisomer (XV) has been formed? Explain why this diastereoisomer is the main product.

This is not an easy question; a paper by Karabatsos[10] presents the hard part of the problem—why this particular diastereoisomer of XIV is formed—and offers an answer to it.

References

1. W. S. Johnson, *Accounts Chem. Res.*, **1**, 1 (1968).
2. E. E. van Tamelen, M. A. Schwartz, E. J. Hessler and A. Storni, *Chem. Comm.*, 409 (1966).
3. W. S. Johnson, K. Wiedhaup, S. F. Brady and G. L. Olson, *J. Amer. Chem. Soc.*, **90**, 5277 (1968).
4. W. S. Johnson, M. F. Semmelhack, M. U. S. Sultanbawa and L. A. Dolak, *J. Amer. Chem. Soc.*, **90**, 2994 (1968).
5. M. Julia, S. Julia, T. S. Yu and C. Neuville, *Bull. Soc. chim. France*, 1381 (1960).
6. S. F. Brady, M. A. Ilton and W. S. Johnson, *J. Amer. Chem. Soc.*, **90**, 2882 (1968); this process is a modification of a method due to Julia.
7. W. S. Johnson, T.-T. Lee, C. A. Harbert, W. R. Bartlett, T. R. Herrin, B. Staskun and D. H. Rich, *J. Amer. Chem. Soc.*, **92**, 4461 (1970).
8. R. E. Ireland, M. I. Dawson and C. A. Lipinski, *Tetrahedron Letters*, 2247 (1970).
9. J. W. Cornforth, R. H. Cornforth and K. K. Mathew, *J. Chem. Soc.*, 112 and 2539 (1959).
10. G. J. Karabatsos, *J. Amer. Chem. Soc.*, **89**, 1367 (1967).

Cecropia Juvenile Hormone

(1)

One of the outstanding features of the brilliant work of E. J. Corey has been the number of new reactions he has devised. Most of these have been developed with a specific aim in mind, usually the synthesis of a natural product. An outstanding example of the development and application of such new reactions is the synthesis[1] of the juvenile hormone (1) of the *Cecropia* beetle. (This compound is potentially useful for the control of various insects). The major problem, as in the case of Johnson's steroid synthesis, p. 156, is the need to obtain only the isomer with the right configuration about each of the double bonds. The Wittig reaction is notorious for being unselective in this respect, although some measure of control can be exercised.[2,3]

Birch reduction of *p*-methoxytoluene gave the diene (2). One equivalent of ozone attacked the more nucleophilic double bond, the intermediate ozonide reacted with dimethyl sulphide, and the aldehydoester was reduced by sodium borohydride to give the hydroxyester (3). The dimethyl sulphide operated by displacing oxygen bonded to oxygen (I or II, for example, and there are other possibilities) to give dimethyl sulphoxide and, in this case, the corresponding aldehydo-ester; this technique[4] for reducing ozonides has proved to be very useful. The result of this sequence was that one double bond had been set up with the desired *cis* configuration retained from the original six-membered ring. Tosylation and lithium aluminium hydride reduction removed the functionality from the ethyl side-chain and at the same time gave the alcohol (4). The tosylate of this alcohol was prepared and the tosylate group displaced by the acetylene anion (5) in hexamethyl phosphoramide solution. The use of aprotic dipolar solvents of this kind[5] has found wide application in recent years, particularly in nucleophilic displacement reactions. It is thought that anions in such solvents are not solvated (although cations *are*) and are therefore much more nucleophilic than they are in protic solvents.

The protecting group was removed with acid to give the acetylenic alcohol (6). The alcohol (6) was reduced with lithium aluminium hydride in the presence of sodium methoxide, and the resulting mixture treated with iodine at $-60°$. This procedure gave stereoselectively the iodoalcohol (7) and had been specially developed by Corey with this kind of application in mind. The mechanism is obscure. The presence of the hydroxyl group in 6 is necessary for the reaction, presumably because it becomes bonded to the aluminium, which then acts as an electrophilic (Lewis acid) centre to assist the attack of hydride on the triple bond. (The reduction of allylic alcohols to saturated alcohols by lithium aluminium hydride is probably similar.) A likely intermediate, then, was the organoaluminium derivative (III),

which underwent electrophilic substitution with iodine (especially since it was a vinyl derivative) with *retention* of configuration. The methoxide was also necessary; in the presence of Lewis acids, and in the absence of the methoxide, the reduction would have taken a different course and an iodoalcohol (IV) of different structure would have been obtained instead of 7.

Corey had also developed a reaction for alkylating vinyl iodides such as 7. Reaction of an iodide with an organocopper reagent, in this case lithium diethylcopper

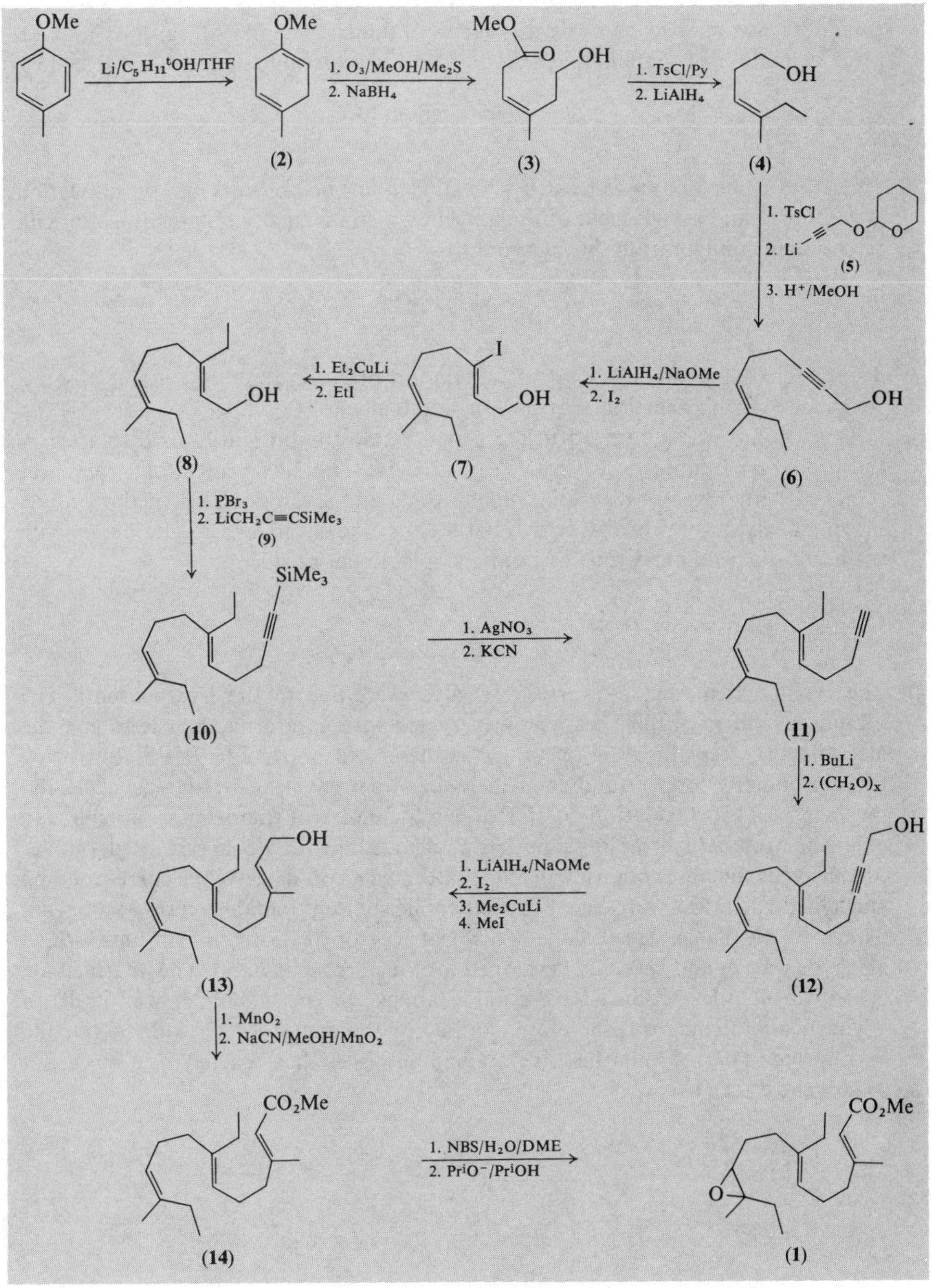
OMe
Li/C5H11tOH/THF
OMe
(2)
1. O3/MeOH/Me2S
2. NaBH4
MeO
O
OH
(3)
1. TsCl/Py
2. LiAlH4
OH
(4)
1. TsCl
2. Li
(5)
3. H+/MeOH
O
O
OH
(6)
1. LiAlH4/NaOMe
2. I2
I
OH
(7)
1. Et2CuLi
2. EtI
OH
(8)
1. PBr3
2. LiCH2C≡CSiMe3
(9)
SiMe3
(10)
1. AgNO3
2. KCN
(11)
1. BuLi
2. (CH2O)x
OH
(12)
1. LiAlH4/NaOMe
2. I2
3. Me2CuLi
4. MeI
OH
(13)
1. MnO2
2. NaCN/MeOH/MnO2
CO2Me
(14)
1. NBS/H2O/DME
2. PriO−/PriOH
CO2Me
O
(1)

(made by mixing two equivalents of ethyl lithium and one of cuprous iodide), gives a mixture of the alkylated product and new organocopper derivatives (V).

$$RI \xrightarrow{\text{Et}_2\text{CuLi}} REt + RCu\}$$
(V)

However, if this mixture is treated with ethyl iodide before work-up, the alkylation is continued and a high yield of the alkylated olefin is finally obtained, again with retention of configuration throughout.

$$RCu\} \xrightarrow{\text{EtI}} REt$$
(V)

The detailed steps involved in this alkylation procedure are at present obscure Applied to **7**, this reaction sequence gave the alcohol (**8**).

This alcohol was converted to the bromide and the bromide was displaced by the propargyl lithium derivative (**9**) to give **10**. The silyl protecting group was removed with silver nitrate followed by potassium cyanide. This method of removing a silyl group[6] probably worked because the silver formed a complex with the acetylene group and thus assisted a displacement at silicon:

$$\overset{\text{H}_2\text{O}}{\underset{\text{Me}_3\text{Si}-\text{C}\equiv\text{C}-}{}} \overset{\text{Ag}^+}{\longrightarrow} \text{Ag}-\text{C}\equiv\text{C}-$$

The cyanide treatment converted the silver acetylide to the hydrocarbon. The lithium derivative of this acetylene was treated with paraformaldehyde to give the alcohol (**12**). Repetition of the sequence described above for **7 → 8**, but using lithium dimethylcopper (and then methyl iodide) gave, again stereospecifically, the alcohol (**13**). Oxidation of this allylic alcohol with manganese dioxide (an oxidising agent which rapidly converts allylic alcohols to α,β-unsaturated ketones or aldehydes, but does not usually affect saturated alcohols) gave the corresponding aldehyde, which was transferred to methanolic sodium cyanide solution and again oxidised with manganese dioxide. This neat way of oxidising an α,β-unsaturated aldehyde was, again, specially developed for the purpose in hand. The intermediate cyanohydrin (VI), produced by cyanide attack on the aldehyde, was itself an allylic alcohol, which was therefore oxidised by the manganese dioxide to give the acyl cyanide (VII). The latter, like an acid chloride, then reacted with methanol to give the ester (**14**).

CHO $\xrightarrow{\text{HCN}}$ (VI) $\xrightarrow{\text{MnO}_2}$ (VII) $\xrightarrow{\text{MeOH}}$ CO₂Me

N-Bromosuccinimide in aqueous dimethoxyethane gave a bromohydrin of **14**, which was closed with alkoxide to give the racemic epoxide of the juvenile hormone (**1**).[7] The preferential attack at this double bond (the yield was 52 per cent) needs comment. The conjugated double bond was obviously less nucleophilic and therefore less readily oxidised than the other two. Both the other double bonds were, however, encumbered with substituents to about the same extent. The selectivity for the first double bond was probably a consequence of the shape the non-polar part of the molecule (**14**) adopts in the polar solvent: the hydrocarbon chain coils up, embedding the central double bond and leaving the other double bond relatively exposed.[8]

The synthesis described above is remarkable, not only for the new reactions developed with particular uses in mind, but also for the newness, in relative terms, of almost every step. The use of dimethyl sulphide, of dipolar aprotic solvents, of lithium aluminium hydride, of silyl protecting groups, of Birch reduction, and of each of the new processes: these are all reactions or reagents which have come into use only recently, most of them since 1965. In 1945, a chemist would have recognised only two or three of the reactions used in this synthesis.

Questions

1. The formation of IV when aluminium chloride/lithium aluminium hydride followed by iodine is used is, as with the formation of III and of **7**, obscure. Suggest a possible explanation.

2. An alternative route from **8** to **12** was also used: the bromide from **8** (RBr) was reacted with the organometallic reagent, phenylthiomethyl copper, $PhSCH_2Cu$,[9] to give the sulphide (VIII). Reaction of this sulphide with methyl iodide and sodium iodide in dimethylformamide gave an iodo compound (IX). The iodide of IX was displaced with the acetylene (**5**) and the protecting group removed. Suggest a mechanism and identify the favourable factors involved in the conversion of VIII to IX.

$$ROH \xrightarrow{} RBr \xrightarrow{PhSCH_2Cu} RCH_2SPh \xrightarrow{MeI/NaI} RCH_2I \xrightarrow[2.H^+]{1.5} (12)$$

$$\text{(8)} \qquad\qquad\qquad \text{(VIII)} \qquad\qquad \text{(IX)}$$

References

1. E. J. Corey, J. A. Katzenellenbogen, N. W. Gilman, S. A. Roman and B. W. Erickson, *J. Amer. Chem. Soc.*, **90**, 5618 (1968).
2. For discussions of this aspect of Wittig reactions see H. O. House, *Modern Synthetic Reactions*, p. 253, Benjamin, New York, 1965, and A. J. Kirby and S. G. Warren, p. 186, *The Organic Chemistry of Phosphorus*, Elsevier, Amsterdam, 1967.
3. E. J. Corey and H. Yamamoto, *J. Amer. Chem. Soc.*, **92**, 226 (1970); M. Schlosser and K. F. Christmann, *Annalen*, **708**, 1 (1967).

4. J. J. Pappas and W. P. Keaveney, *Tetrahedron Letters*, 4273 (1966).
5. A. J. Parker in *Advances in Organic Chemistry, Methods and Results*, Vol. 5, Interscience, New York, 1965.
6. H. M. Schmidt and J. F. Arens, *Rec. Trav. Chim.*, **86**, 1138 (1967).
7. The absolute configuration of the natural hormone has been shown to be 10R, 11S, D. J. Faulkner and M. R. Petersen, *J. Amer. Chem. Soc.*, 3766, **93** (1971), and K. Nakanishi, D. A. Schooley, M. Koreeda and J. Dillon, *Chem. Comm.*, 1235 (1971).
8. E. E. van Tamelen and T. J. Curphey, *Tetrahedron Letters*, 121 (1962).
9. This reagent and the sequence R—Br → RCH$_2$I, were developed by E. J. Corey and M. Jautelat, *Tetrahedron Letters*, 5787 (1968).

A Second Synthesis of the *Cecropia* Juvenile Hormone

(1)

In addition to Corey's synthesis, there have been several other successful and attractive syntheses of the juvenile hormone (**1**). One of these,[1] described here, tackled the problem of generating the double bonds stereospecifically by using the fragmentation of a bicyclic precursor. Control of the olefin geometry is then made a matter of control of relative stereochemistry in the cyclic system, just as it was in Corey's synthesis of caryophyllene (p. 137). This approach is the reverse of that used by W. S. Johnson (see p. 156) in his work on olefin cyclisation, where the double bonds were set up stereospecifically, and stereospecifically cyclised to polycyclic compounds. In setting up the bicyclic system in this synthesis, several reactions were used, the stereochemical outcome of which was easily predictable from analogies in the steroid field.

The bicyclic system was prepared by a Robinson ring extension sequence using 2-ethylcyclopentan-1,3-dione (**3**) and propyl vinyl ketone (**2**). Selective reduction of the product (**4**) gave **5**, in which only the cyclopentanone carbonyl group had been reduced. There are many analogies for such selectivity and stereoselectivity in the steroid field;[2] for example, steroidal-4-ene-3,17-diones can readily be reduced with sodium borohydride to give the corresponding 17β-hydroxy-4-ene-3-one. The hydroxyl group of **5** was protected, and alkylation of the enolate of the ketone then gave the unconjugated ketone (**6**). Again there was ample analogy for enolisation in this direction, rather than the other, and likewise for alkylation at this site and this less hindered side of the molecule; there are examples in the Woodward synthesis of steroids (p. 61) and in the Stork synthesis of dehydroabietic acid (p. 77). Acid hydrolysis and reduction then gave the diol (**7**). The reducing agent, lithium tri-t-butoxyaluminium hydride, is made from lithium aluminium hydride and three equivalents of t-butanol, and is one of several modified aluminium hydride reducing agents. This one shows enhanced stereoselectivity in comparison with the unmodified reagent. For example, it is well known that cholestanone gives predominantly (~90%) the equatorial (i.e. 3β) alcohol with lithium aluminium hydride; with the modified reagent, the proportion of 3β-alcohol is 99 per cent.

The configuration of the new hydroxyl group in 7 was demonstrated by the addition of hypobromous acid to the diacetate (I), followed by reductive removal of the bromine, to give the hydroxydiacetate (II). The n.m.r. spectrum of II showed that H-5 appeared at unusually low field as a result of the spatial proximity of the α-hydroxy group on C-3 (indane numbering).

Epoxidation of the diol (7) in ether gave the α-epoxide (8), which was not the one wanted. The configuration of this epoxide was discovered because the same α-epoxide was obtained by treatment of the bromohydrin (I) with alkali. Fortunately, epoxidation in methylene dichloride gave largely the β-epoxide (9), as a result of the influence of the hydroxyl group. It was well known, from work by Henbest,[3] that hydroxyl groups can direct the delivery of the oxygen atom of peracids to the double bond; this evidently works in methylene dichloride, but not in ether. Lithium aluminium hydride reduction of the epoxide gave the triol (10), the reducing agent having, as usual, delivered hydride ion to the less substituted end of the epoxide. The C-5 hydroxyl group of 10 was the least hindered, and was the only one to become tosylated under mild conditions. The tosylate (11) was now ready for the first fragmentation reaction: treatment with sodium hydride gave the olefin (12) quantitatively. The reaction was stereospecific and easy because the orbitals involved were well arranged for continuous overlap (III), in the ∖⁄∖⁄ arrangement well known from Grob's work[4] to be the most favourable for concerted fragmentation.

(III)

The hydroxyl group of 12 was temporarily protected and the ketone treated with methyl lithium to give the diol (13). Tosylation of 13 gave the monotosylate (14), tertiary hydroxyl groups being generally very difficult to tosylate. The second fragmentation again proceeded smoothly, using sodium hydride, to give the *trans,cis*-6-ethyl-10-methyldodeca-5,9-dien-2-one (15). The configuration of

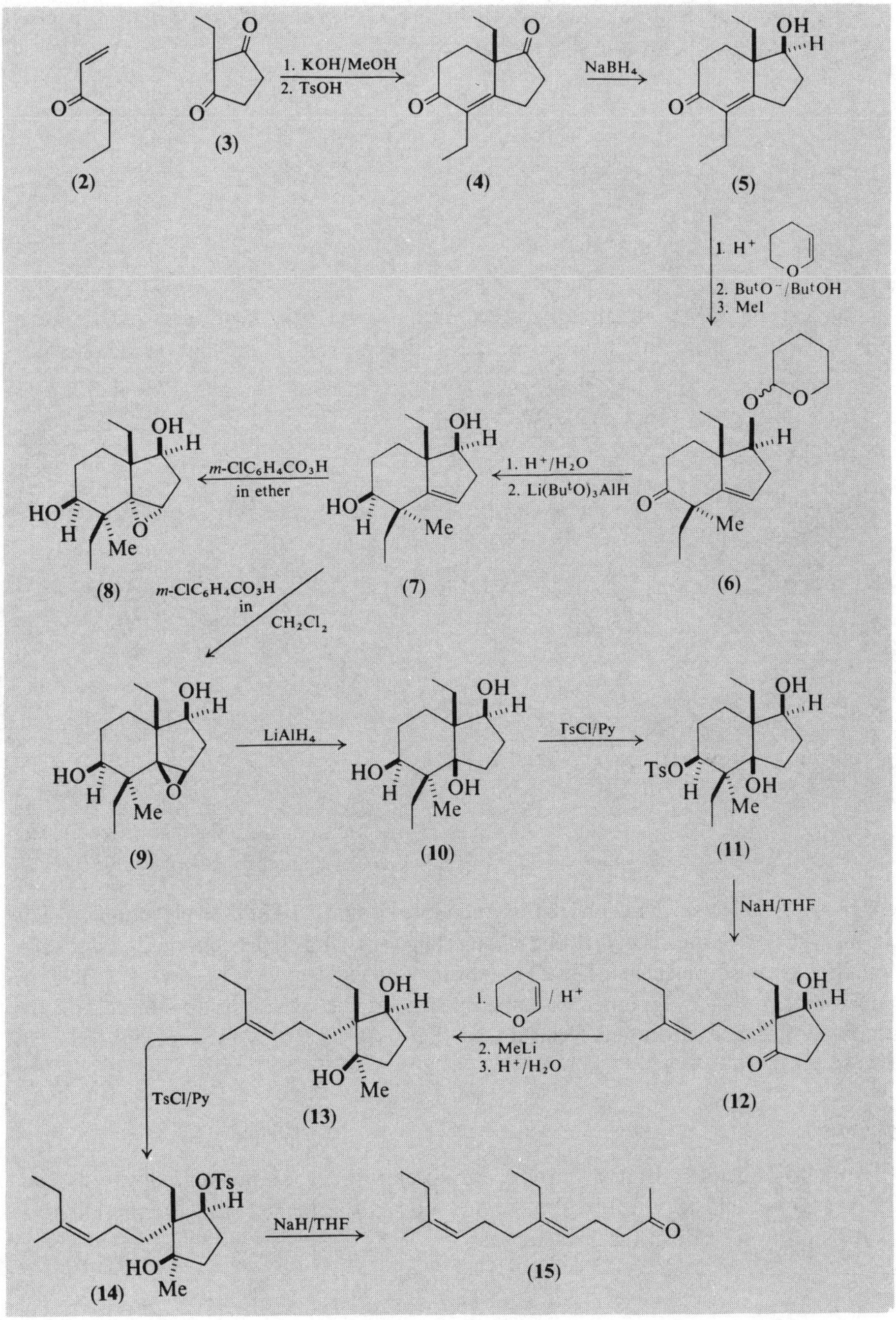
(2)
(3)
1. KOH/MeOH
2. TsOH
(4)
NaBH₄
(5)
1. H⁺
2. ButO⁻/ButOH
3. MeI
(6)
1. H⁺/H₂O
2. Li(ButO)₃AlH
(7)
m-ClC₆H₄CO₃H in ether
(8)
m-ClC₆H₄CO₃H in CH₂Cl₂
(9)
LiAlH₄
(10)
TsCl/Py
(11)
NaH/THF
(12)
1. / H⁺
2. MeLi
3. H⁺/H₂O
(13)
TsCl/Py
(14)
NaH/THF
(15)

the second double bond followed from the *cis* arrangement (IV) of the ethyl and tosyloxy groups on the five membered ring.

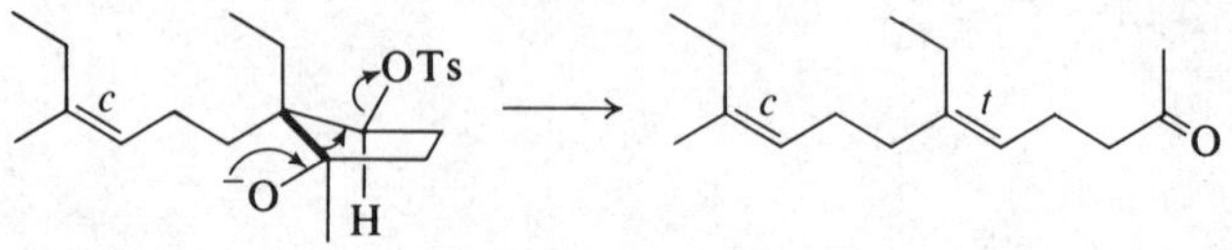

(IV)

The ketone (**15**) had already been used in the first synthesis[5] of the juvenile hormone. This first synthesis had suffered from a lack of stereochemical

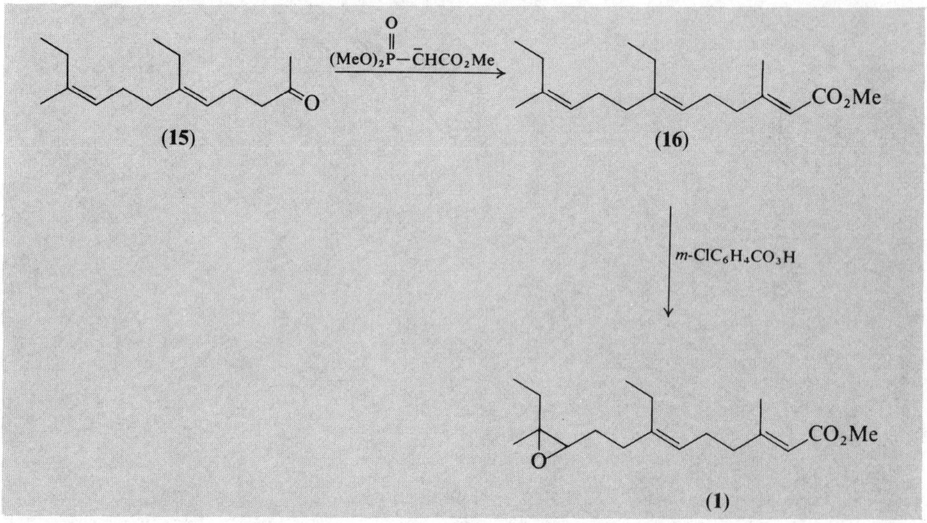

control— a deficiency which both this and Corey's synthesis effectively made up. The ketone (**15**) was combined with the Wittig reagent giving 30 per cent of the ester (**16**), together with 18 per cent of the *cis* isomer. Epoxidation of **16** gave the juvenile hormone (**1**) (40%) (together with the epoxide at the central double bond and the diepoxide) in a reaction which paralleled that used by Corey for the conversion of **16** (**14** on p. 167) to **1**.

Question

How do you think that a hydroxy group can direct the delivery of the oxygen atom of a peracid to a double bond, and how does this account for the effect of changing the solvent in the present case?

References

1. R. Zurflüh, E. N. Wall, J. B. Siddall and J. A. Edwards, *J. Amer. Chem. Soc.*, **90**, 6224 (1968).
2. A good source of analogies such as this one is: C. Djerassi (Ed.), *Steroid Reactions*, Holden-Day, San Francisco, 1963.
3. H. B. Henbest, Special Publication of the Chemical Society, No. 19, *Organic Reaction Mechanisms*, London, 1965, p. 83.
4. C. A. Grob and P. W. Schiess, *Angew. Chem. Internat. Edn.*, **6**, 1 (1967).
5. K. H. Dahm, B. M. Trost and H. Röller, *J. Amer. Chem. Soc.*, **89**, 5292 (1967).

Lycopodine

(1)

Lycopodine (**1**) is a widely distributed alkaloid. Its synthesis presented many difficulties to the several groups who worked on the problem in the 1960's. The routes they developed[1-5] are strikingly different from one another, particularly with regard to the order in which the bonds to C-13 are put in. A high degree of stereochemical control, achieved by good design, distinguishes the synthesis[5] described in this chapter.

Reaction of ethyl acrylate with triphenylphosphine and *m*-methoxybenzaldehyde gave, in one step, 4-*m*-methoxyphenylbut-3-enoate (**2**). This compound with ethyl acetoacetate and sodium ethoxide, followed by hydrolysis and decarboxylation, gave the cyclohexandione (**3**). Presumably the double bond of **2**, under the influence of the base, became conjugated to the carbethoxy group, and the subsequent steps involved Michael and Claisen processes. The β-diketone (**3**) gave a monoenol ether which was reduced with lithium aluminium hydride to give, after acidic work-up, the unsaturated ketone (**4**). Methyl magnesium iodide, with a catalytic amount of cuprous chloride, gave the saturated ketone (**5**). There are two points to be noted here. First, the use of cuprous ion (a methylcopper reagent being a probable intermediate) ensures the conjugate addition—in its absence attack is usually more rapid at the carbonyl group itself. Secondly, the reaction is stereo-selective for reasons which require rather careful explanation. In order to maintain the continuous overlap of orbitals, the attack of the methylcopper reagent on the unsaturated ketone must take place from a vertical—that is an axial—direction. The most probable conformation of the unsaturated ketone is shown in I and again

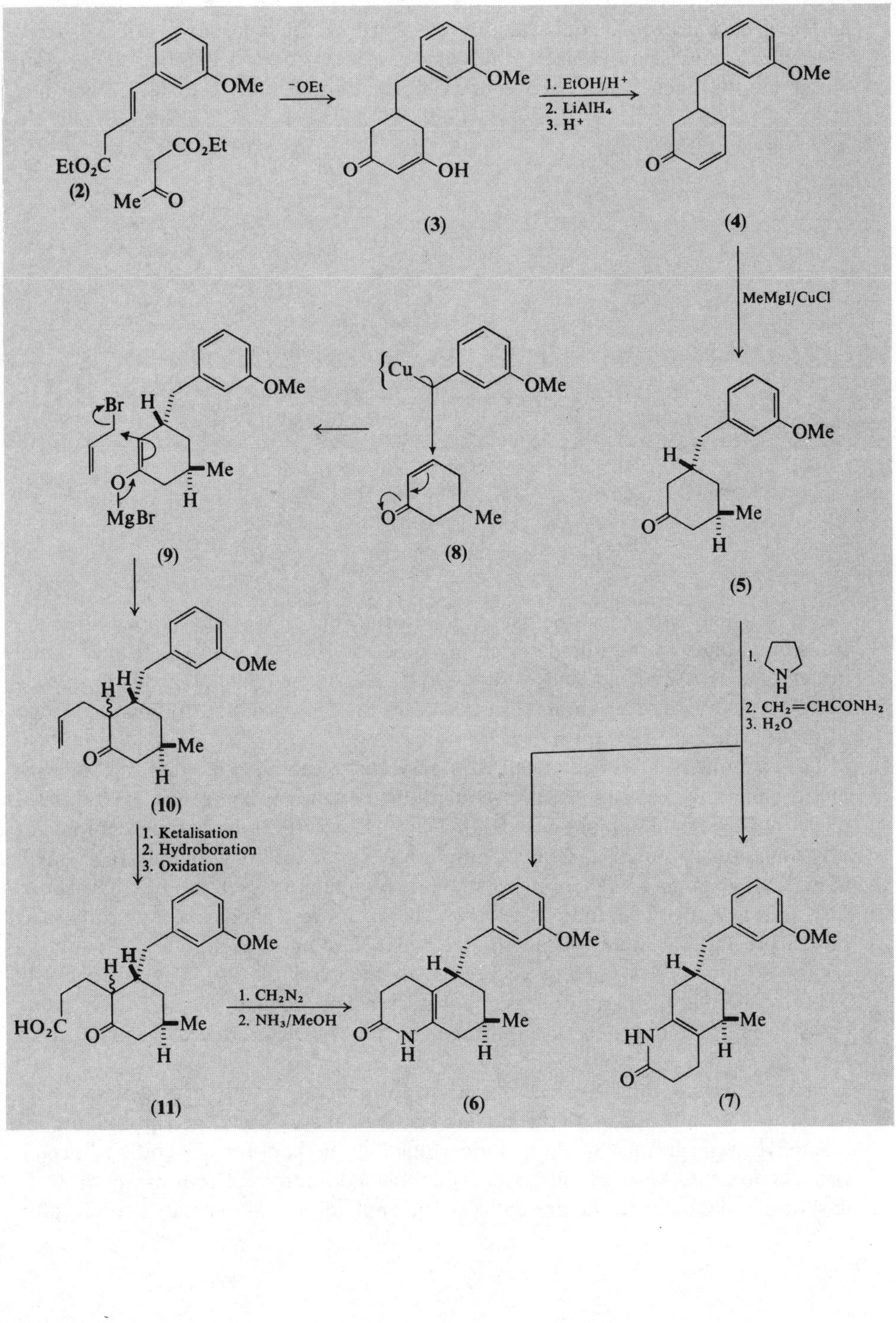
OMe
EtO2C
(2)
CO2Et
Me
O
-OEt
OMe
O
OH
(3)
1. EtOH/H+
2. LiAlH4
3. H+
OMe
O
(4)
MeMgI/CuCl
OMe
H
Me
O
H
(5)
Cu
OMe
O
Me
(8)
OMe
Br
H
Me
O
MgBr
H
(9)
OMe
H
H
Me
O
H
(10)
1. Ketalisation
2. Hydroboration
3. Oxidation
OMe
H
H
HO2C
O
Me
H
(11)
1. CH2N2
2. NH3/MeOH
1.
N
H
2. CH2=CHCONH2
3. H2O
OMe
H
O
N
H
Me
H
(6)
OMe
H
HN
Me
H
O
(7)

in III; in each drawing, axial attack is illustrated. In the latter, axial attack is from below and leads to an enolate which must, at first, adopt a boatlike shape (IV). Later this product would change its conformation to the more stable chair-like

shape in which both substituents are equatorial, but this would happen after the transition state for the rate-determining process (III → IV). In the former, axial attack is from above and leads directly to a chair-like enolate (II); this process is therefore favoured, even though, in this case, the final product (5) must have one of the substituents axial on the cyclohexane ring.

The enamine of **5** was alkylated with acrylamide to give a mixture of the hexa-hydro quinolones (**6**) and (**7**). I have mentioned enamines before (p. 75) as useful nucleophilic carbon compounds. In this case the electrophilic carbon component was an $\alpha\beta$-unsaturated carbonyl compound; this class of compounds is much used with enamines.[6] The intermediate propionamide cyclised in an obvious way. Since the two enamines of **5** were likely to be very similar energetically, it is not surprising that there was a product corresponding to each. **6** and **7** could be separated, and **6** was used to proceed with the synthesis. But it was obviously desirable, for aesthetic as well as practical reasons, to find an alternative route which would avoid a step as unselective as that which produced the mixture of **6** and **7**.

The alternative route involved a method for generating only that enolate which would lead to alkylation at the carbon between the carbonyl group and the *m*-methoxybenzyl group. Stork had developed a method for generating *specific enolates* for this kind of purpose, using the addition of electrons to an α,β unsaturated ketone.[7] In the present case the enolate (**9**) was generated specifically by adding, not electrons, but the *m*-methoxybenzyl group (from the Grignard with cupric chloride) to the unsaturated ketone (**8**). By reaction with allyl bromide this enolate was trapped to give **10**. For the same reason as before, the Grignard addition was stereoselective; the stereochemistry of the alkylation was not important because this chiral centre was lost in the subsequent transformation of the vinyl group: hydroboration, oxidation, and lactam formation gave the quinolone (**6**), which was identical to that obtained before.

Now came the step which converted this molecule (**6**) into something which began to look more like lycopodine. On treatment with acid, **6** gave two products, **12** (55%) and **13** (29%). Protonation of **6** at the carbon marked * in formula V could, and very likely did, give the two possible immonium cations: one with the proton β and one with the proton α. The products obtained, **12** and **13**, were both the result of β-protonation (as shown by the conversion of **12** into lycopodine), a result which had been expected for this reason: in order for cyclisation to take place, a conformation like V is necessary for the β-protonated intermediate, and a

(V) (VI)

conformation like VI would be necessary for the α-protonated intermediate if cyclisation were to take place. The difference between V and VI is that in the latter the propionamide group is axial on the carbocyclic ring. while in the former it is equatorial. The conformation V is, therefore, more often attained, and cyclisation follows from it.

Lithium aluminium hydride reduction, and then Birch reduction of **12** gave the expected product **14**. Hydrolysis, however, did not give, and could not be induced to give, the usual conjugated ketone. Since this unexceptionable approach towards cleavage of the ring was not possible, an alternative pathway was developed which took advantage of the recalcitrance of the fully substituted double bond. Treatment of **14** with a very strong base gave the conjugated diene, in which the other double bond had migrated; the diene was then acylated to give **15**. One double bond was selectively cleaved by ozone to give **16**. This selectivity stemmed from the greater accessibility of the less substituted double bond and from the greater nucleo-philicity of a double bond which is directly bonded to an oxygen atom. **16** could be oxidised with selenium dioxide and hydrogen peroxide to give the Baeyer-Villiger product (17). This reagent often gives carboxylic acids from α,β-unsaturated aldehydes, but in this case it seems likely that an intermediate, such as VII, reacted to give the fully substituted carbonium ion (VIII) rather than lose the proton (H*) bonded to carbon.

(VII)   (VIII)

Methanolysis of 17 gave the corresponding keto-ester, and zinc reduction removed the trichloroethoxycarbonyl group. The amino ester cyclised spontaneously to give the keto-lactam (18). Lithium aluminium hydride reduction followed by reoxidation of the alcohol gave racemic lycopodine.

It is particularly instructive in this synthesis to work backwards through the steps which were finally used, in order to see how the four rings of the tetracyclic system were set up. In this way it is easier to see how the apparently miraculous choice of a substituted benzylcyclohexanone starting material came to be made.

(IX)   (X)

For example, we can transform (see p. 4) the C-12–C-13 bond (IX → X) and recognise that this dislocation reveals those sufficient features for reaction: a nucleophilic and an electrophilic carbon atom. It is sensible dislocations of this kind which make it possible, when planning a synthesis, to choose in broad outline the nucleophilic and electrophilic components needed. It is particularly instructive to look not only at this synthesis but also at the others in the lycopodine field,[1–4] and see how they too have evolved from dislocations of the lycopodine molecule.

In this synthesis, as with most syntheses, a great many routes were followed which did not turn out to lead in the right direction. One early pathway was mentioned above, partly to emphasise how much the synthesis was improved by using a route to 6 alone, instead of one giving a mixture of 6 and 7, and partly to draw attention to the way in which syntheses may develop as they are being

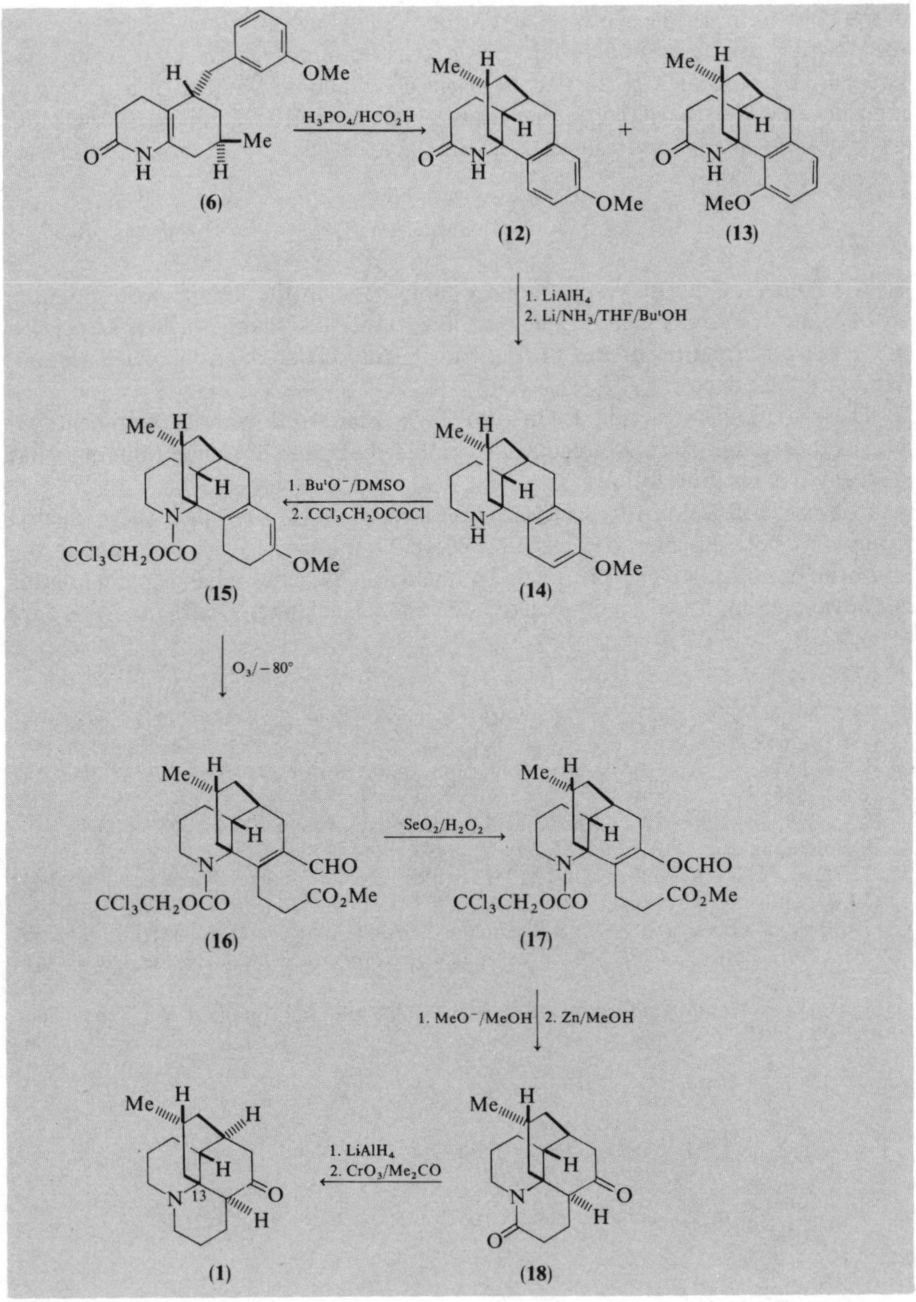

OMe
H
Me
O
N
H
H
(6)
H_3PO_4/HCO_2H
Me
H
H
O
N
H
OMe
(12)
+
Me
H
H
O
N
H
MeO
(13)
1. $LiAlH_4$
2. $Li/NH_3/THF/Bu^tOH$
Me
H
H
N
CCl_3CH_2OCO
OMe
(15)
1. $Bu^tO^-/DMSO$
2. CCl_3CH_2OCOCl
Me
H
H
N
H
OMe
(14)
$O_3/-80°$
Me
H
H
N
CCl_3CH_2OCO
CHO
CO_2Me
(16)
SeO_2/H_2O_2
Me
H
H
N
CCl_3CH_2OCO
OCHO
CO_2Me
(17)
1. $MeO^-/MeOH$
2. Zn/MeOH
Me
H
H
H
N
13
O
H
(1)
1. $LiAlH_4$
2. CrO_3/Me_2CO
Me
H
H
N
O
O
H
(18)

carried out. Some of the syntheses in this book may have given the impression that the practitioners never put a foot wrong. This is, of course, very far from what happens. The use of a particular sequence of reactions, or even of a particular reagent, may *look* inevitable as one reads it, but the brief description of the route or conditions finally used usually conceals the existence of notebooks and bottles full of early failures.

Questions

1. The failure of the fully substituted double bond in the ketone corresponding to **14** to move into conjugation must be due to thermodynamic factors associated with the conformation of the ketone and its (unobtainable) conjugated isomer. What are these factors?

2. The remarkable reaction leading to **2** was described without comment. So brief a description should already have hinted that you should be thinking what pathway this reaction follows.

3. In a model reaction for the cyclisation step (**6 → 12**), in which the *C*-methyl group was not present, no cyclisation could be induced to occur. Examine the conformations available to protonated **6** (including V), and find an explanation for the advantage conferred on **6**, relative to the model lacking the extra methyl group.

References

1. K. Wiesner, Z. Valenta and co-workers, *Tetrahedron Letters*, 1279 (1965); 4931 (1967); 5643 (1968).
2. W. A. Ayer, W. R. Bowman, T. C. Joseph and P. Smith, *J. Amer. Chem. Soc.*, **90**, 1648 (1968).
3. E. Colvin, J. Martin, W. Parker and R. A. Raphael, *Chem. Comm.*, 596 (1966).
4. F. Bohlmann and O. Schmidt, *Chem. Ber.*, **97**, 1354 (1964).
5. G. Stork, R. A. Kretchmer and R. H. Schlessinger, *J. Amer. Chem. Soc.*, **90**, 1647 (1968); G. Stork, *Pure and Appl. Chem.*, **17**, 383 (1968).
6. G. Stork, A. Brizzolara, H. Landesman, J. Szmuszkovicz and R. Terrell, *J. Amer. Chem. Soc.*, **85**, 207 (1963); A. G. Cook (Ed.), *Enamines*, Marcel Dekker, New York, 1969.
7. G. Stork, P. Rosen, N. Goldman, R. V. Coombs and J. Tsuji, *J. Amer. Chem. Soc.*, **87**, 275 (1965).

Colchicine (First synthesis)

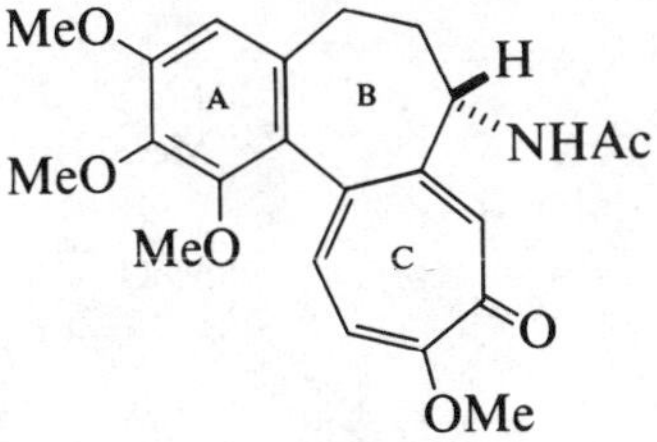

(1)

The alkaloid colchicine, used in the treatment of gout and, because it inhibits mitosis, in biological research, has been the subject of an unusually large number of syntheses. These syntheses are different enough from each other to illustrate the different ways a molecule as complicated as this can be approached. Not that it is all that complicated, having only one, easily controlled, chiral centre and two aromatic rings; but the difficulties encountered in synthetic work in several laboratories would seem to suggest that the particular combination of groupings and the presence of two seven-membered rings (including a tropolone ring, for which there are few general methods of synthesis) do pose rather special problems.

We shall now take this opportunity to look at and compare four of the successful syntheses; the first is the one devised by Eschenmoser. These four syntheses do not exhaust the repertoire: two other complete syntheses have been reported and there are many reports of uncompleted syntheses.

Eschenmoser's synthesis[1] began with purpurogallin trimethylether (3). This well known compound,[2] produced by the oxidation of pyrogallol (2), followed by methylation, already has the carbon skeleton of the A and B rings. When 3 was hydrogenated, reduced by lithium aluminium hydride, and treated with phosphoric acid, it gave the benzsuberone (4). In this sequence, the hydrogenation removed the C=C double bonds and the lithium aluminium hydride reduced the ketone (I) to give the alcohol (II). The phosphoric acid then caused the dehydration of the benzylic alcohol, and, since the product (III) was an enol ether, it readily hydrolysed.

The next reaction introduced three more carbon atoms. The benzsuberone (4) was alkylated in a base-catalysed Michael reaction with methyl propiolate; the product isolated was the 2-pyrone (5). The sequence of reactions which took place in the course of this transformation may be inferred from some of the intermediates isolated when milder conditions were used. At room temperature with

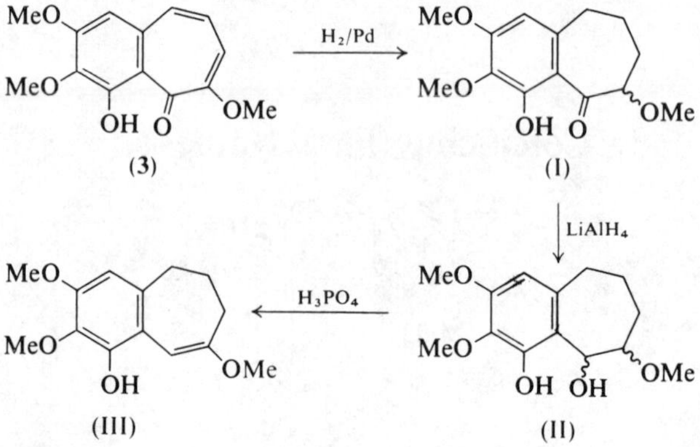

(3) → (I) → (II) → (III)

0·1 equivalents of base, the acrylic ester (V) was obtained; on being warmed, this was converted to the saturated derivative (VII). Further treatment with one equivalent of base caused the elimination of the phenolate ion to give IX. We

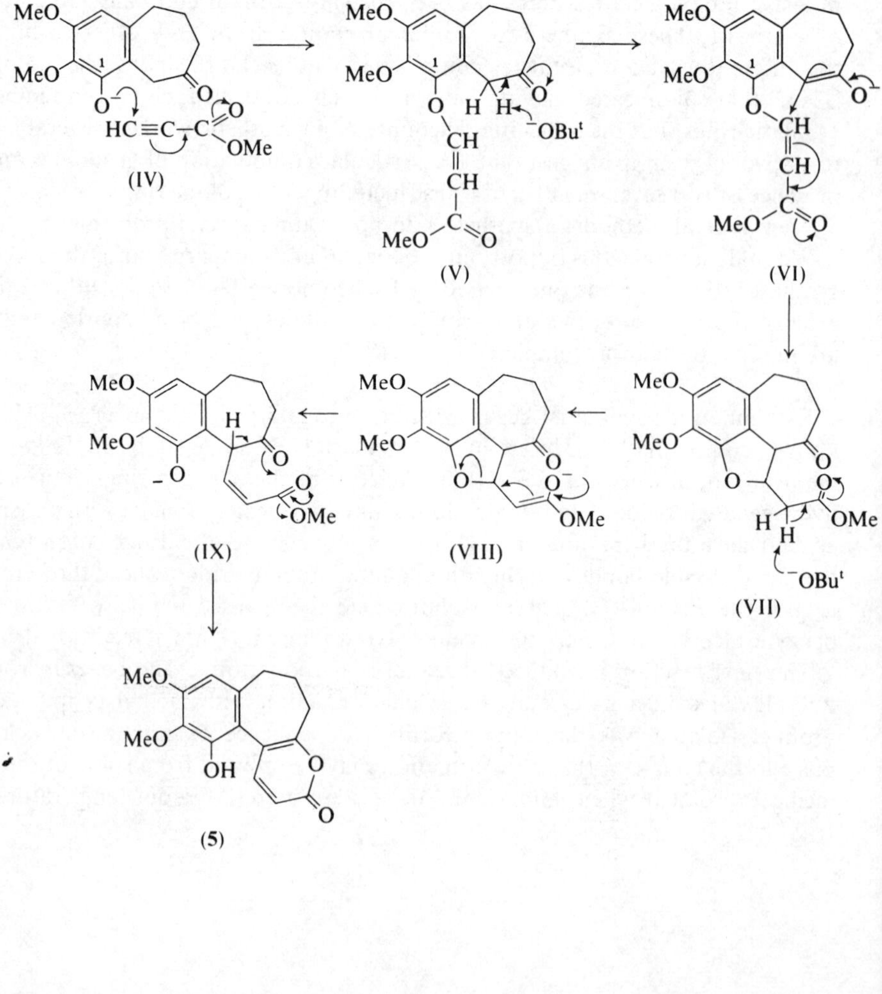

(IV) (V) (VI)

(IX) (VIII) (VII)

(5)

can now see the reason for having the 1-hydroxyl group free in this sequence—
it served as a platform from which to deliver the three-carbon unit. It therefore
ensured that the product was the one derived from enolisation towards the benzene
ring (VI) and not the one from the other possible enolate. Enolisation towards
the ring is favoured in any case, but alkylation of this enolate is sterically more
hindered than alkylation of the other enolate. We shall see, in the next synthesis
to be described, that it was possible to alkylate at this position without using the
1-hydroxyl as a platform.

Having served its purpose, the 1-hydroxyl group was methylated and the pyrone
was then combined in a Diels-Alder reaction with chloromethylmaleic anhydride
to give, mainly, one (**6** = X) of the two possible adducts (X and XI). Diels-Alder

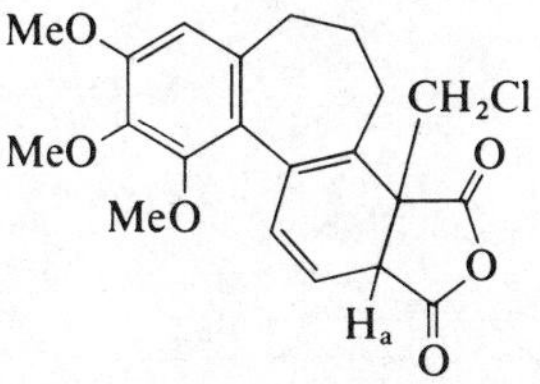

(X)

adducts to pyrones do not always lose the carbon dioxide from the first-formed
product, but under the fairly vigorous conditions of this reaction it is not surprising
that the intermediate is not isolated. As it happens, it is not necessary that the
adduct should be this one (**6** = X) rather than the other (XI), but it is always

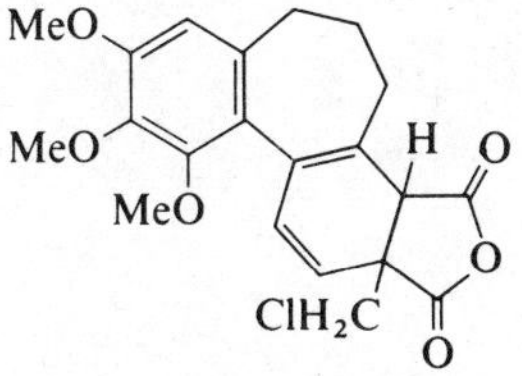

(XI)

agreeable to be able to say which is which. In this case, the n.m.r. spectrum
showed a signal at $\tau 6.64$, which could be assigned to the allylic hydrogen next to
the anhydride group (H_a in X); this signal was a doublet ($J \approx 11$ Hz), whereas
it would have been a singlet if the structure had been XI.

The adduct (6) was converted to its dimethyl ester (7), which was treated with potassium t-butoxide. In the ensuing reaction, the enolate ion (XII) displaced the chlorine and gave a norcaradiene derivative (XIII), which opened, in a well known

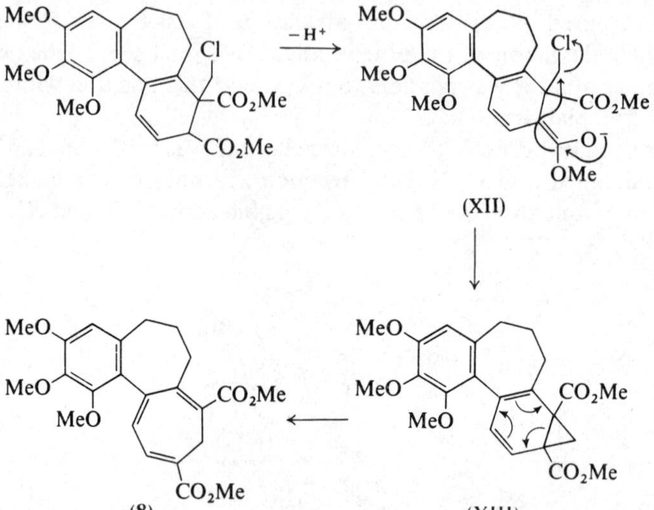

(XII)

(8)   (XIII)

electrocyclic reaction, to give the cycloheptatriene diester (8). It was possible that this reaction would give the diester (XV), by a sequence (XII → XIV → XV),

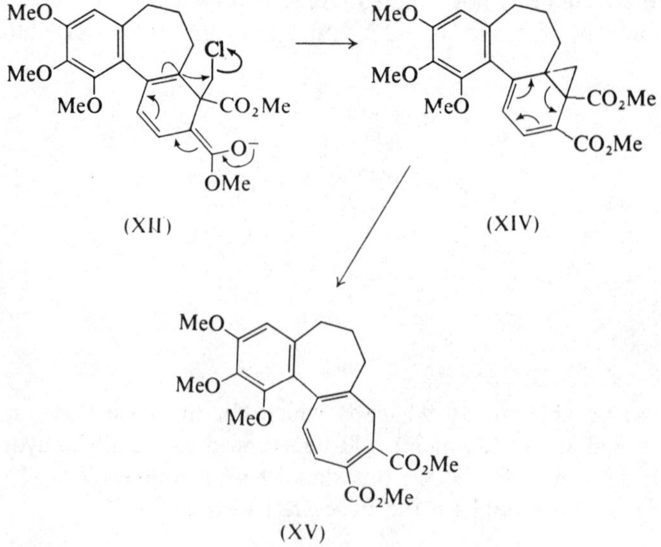

(XII)   (XIV)

(XV)

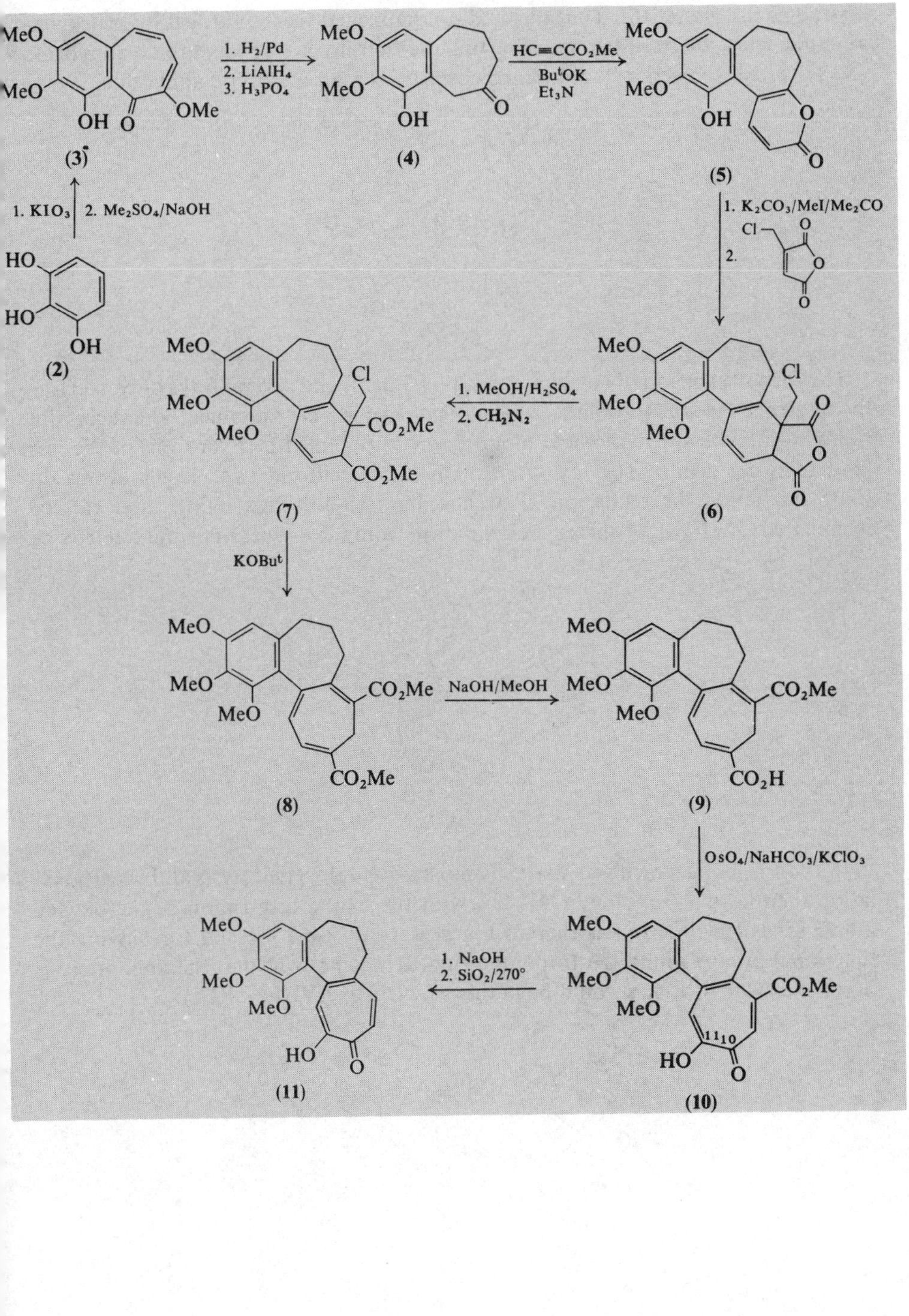

instead of the diester (**8**). That this had not happened was confirmed by hydrogenating the triene diester (**8**) and converting the (tetrahydro) product to an anhydride (XVI), which showed infrared bands characteristic of a glutaric anhydride.

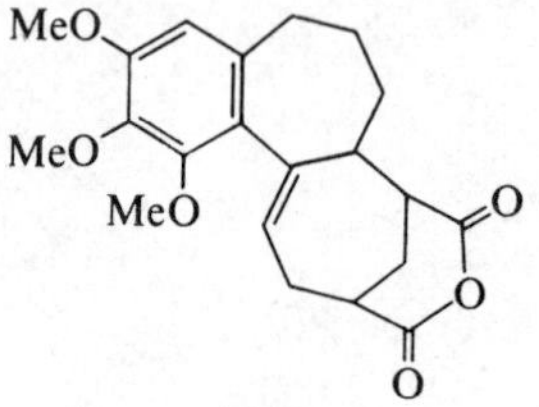

(XVI)

The carbon skeleton of colchicine was now intact, and several ways of converting the diester to give the tropolone system of colchicine were examined. The successful route began with the mono-acid mono-ester (**9**) obtained by hydrolysis of the less hindered ester group of **8**. Oxidation with osmic acid put two oxygen atoms on C-10 and C-11; the oxidation also caused mono-decarboxylation, and can be formulated (XVII) as a concerted elimination from the adduct of osmium tetroxide

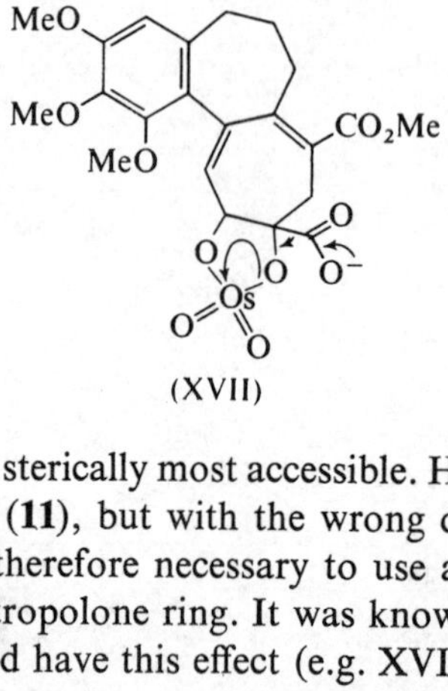

(XVII)

to the double bond which was sterically most accessible. Hydrolysis and decarboxylation then gave a tropolone (**11**), but with the wrong disposition of the oxygen atoms for colchicine. It was therefore necessary to use a method for moving the functional groups round the tropolone ring. It was known[3] that the ammonolysis of an α-tosyloxytropone would have this effect (e.g. XVIII → XIX):

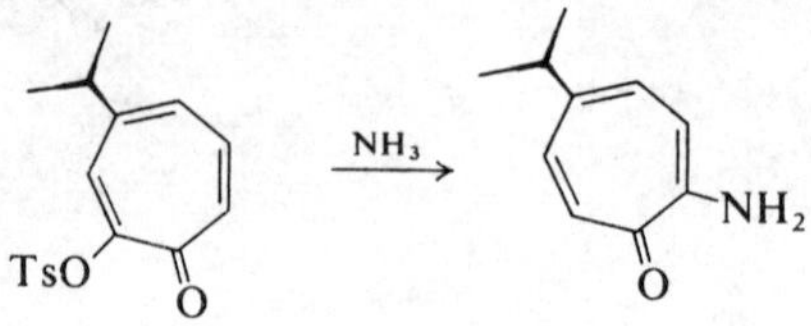

(XVIII)　　　　　　　(XIX)

This method was therefore applied: tosylation of the tropolone (11) gave a mixture of tosylates (XX and XXI), and ammonolysis of this mixture gave a mixture of

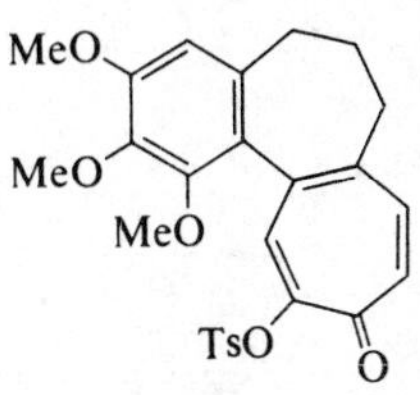

(XX) 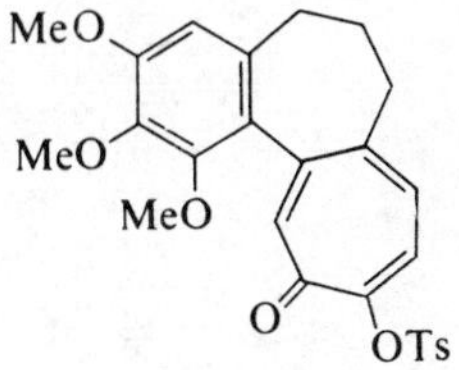 (XXI)

products from which 12 crystallised. This product could be recognised as being 12, and not one of the other possible isomers, by its ultraviolet spectrum, which was identical to that of colchicinamide (XXII), the corresponding compound derived

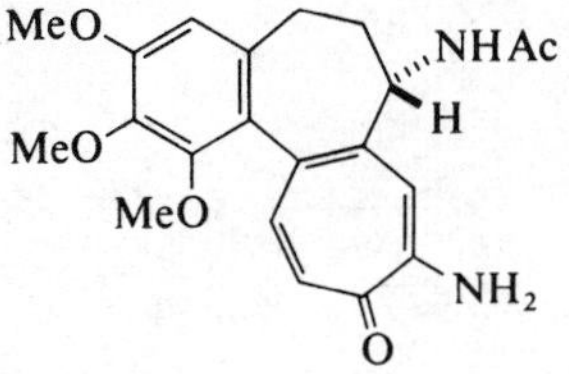

(XXII)

from colchicine. Alkaline hydrolysis replaced the amino group with a hydroxyl to give desacetylamino colchiceine* (13), identical to material obtained by degradation of colchicine. In this sequence of reactions, nucleophilic attack on the

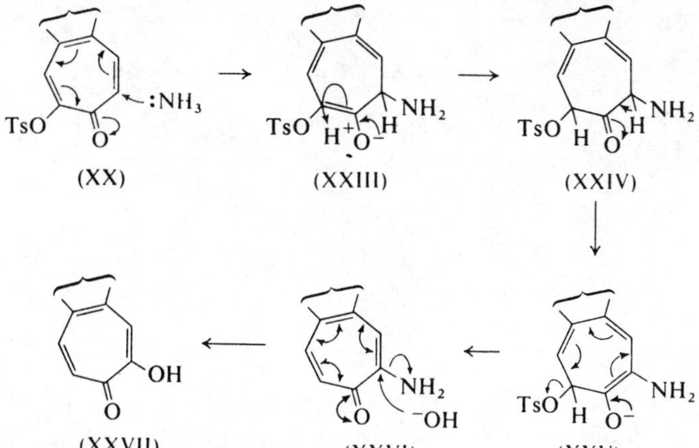

(XX) (XXIII) (XXIV)

(XXVII) (XXVI) (XXV)

* The name colchiceine, as distinct from colchicine, is used for the tropolones, such as 13 and 15, which do not have the methyl ether group present.

(11) → (12) → KOH → (13) → (14) → (15) → (1)

tropone ring occurred, first by ammonia on the tosyloxy tropone (XX). This was followed by ketonisation (XXIII) and enolisation (XXIV), before the tosylate group was expelled (XXV). In the second step, the ammonia was expelled in a simple sequence (XXVI, arrows), a vinylogous version of amide hydrolysis.

From this point on a relay was established: in other words, material obtained by degradation of colchicine was used. Since no chiral centres were present in this degradation product, there had been no need for resolution up to this point. Methylation with diazomethane gave a mixture of both methyl ethers (XXVIII and XXIX). The bromination of the former ether gave a mixture of products

(XXVIII)   (XXIX)

from which the required one (14) could be separated. The other methyl ether (XXIX), which had the methyl group on the right oxygen for colchicine, was not so well set up for bromination at this position: a radical (or cationic) centre at C-7 in this derivative is not conjugated with the benzene ring in the way that the corresponding radical in XXVIII is.

The bromine in 14 was displaced with ammonia, which also caused ammonolysis in the tropone ring. Alkali reversed this process, and the product was then racemic desacetylcolchiceine (15). Methylation with diazomethane again gave both ethers (corresponding to XXVIII and XXIX); acetylation of the one which was desacetyl colchicine gave racemic colchicine. The other ether could be recycled by ammonolysis and treatment with alkali, and was therefore not wasted. The racemic desacetyl colchiceine (15) had been resolved earlier;[4] so the synthesis was complete.

Questions

1. The oxidation of pyrogallol to purpurogallin, although it is a well-known reaction, is not a well-understood reaction. What could the mechanism be?[5]
2. In the ammonolysis of the mixture of tosylates (XX and XXI) there were a number of other products, which were not isolated. What are the other products likely to have been?

References

1. J. Schreiber, W. Leimgruber, M. Pesaro, P. Schudel, T. Threlfall and A. Eschenmoser, *Helv. Chim. Acta*, **44**, 540 (1961).

2. J. A. Barltrop and J. S. Nicholson, *J. Chem. Soc.*, 116 (1948); R. D. Haworth, B. P. Moore and P. L. Pauson, *J. Chem. Soc.*, 1045 (1948).
3. For a review of tropone and tropolone chemistry, see: T. Nozoe in D. Ginsburg (Ed.) *Non-Benzenoid Aromatic Compounds*, Interscience, New York, 1959, p. 339.
4. H. Corrodi and E. Hardegger, *Helv. Chim. Acta*, **40**, 193 (1957).
5. Some discussion of the mechanism of this reaction is given by L. Horner, K. H. Weber and W. Dürckheimer, *Chem. Ber.*, **94**, 2881 (1961).

Colchicine (Second Synthesis)

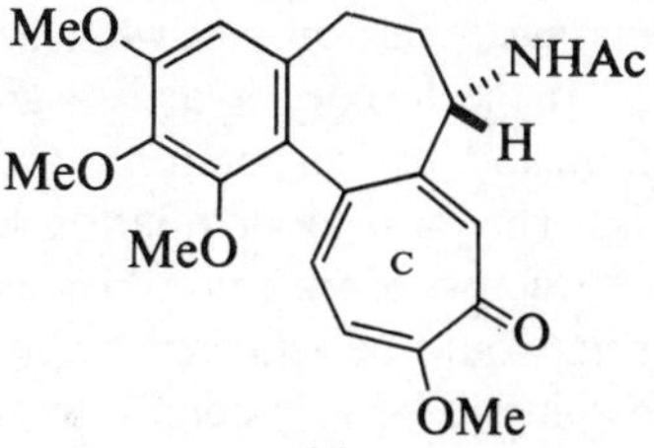

(1)

The second synthesis, by van Tamelen,[1] was carried out at the same time as Eschenmoser's, and followed a similar general plan. It, too, began with purpurogallin, which was completely methylated, and reduced—in a sequence similar to that used by Eschenmoser—to give the benzsuberone (3). In spite of a large number of attempts, only one procedure for the alkylation of 3 was found to work, namely cyanoethylation catalysed by potassium t-butoxide. The product (4) was the result expected from enolisation towards the benzene ring. But, although the corresponding step in Eschenmoser's synthesis was a clear-cut case, the method of synthesis in this case was not unambiguous: it was necessary to prove the structure. Hydrolysis and treatment with acetic anhydride gave the enol lactone (I).

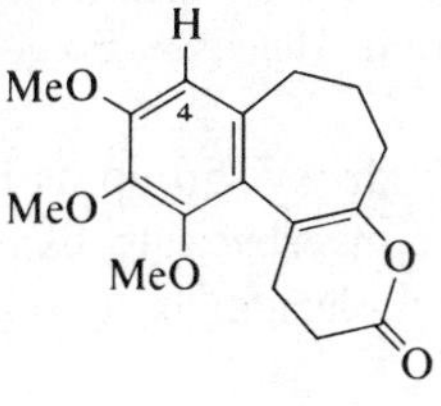

(I) 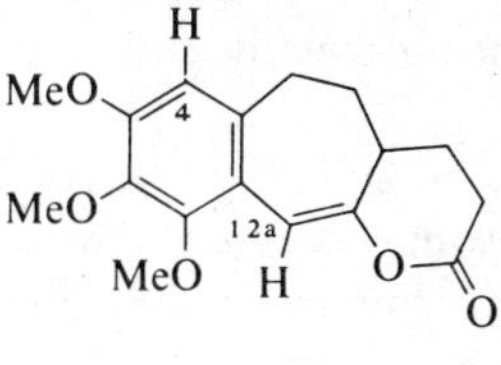　　　　　　　　(II)

The ultraviolet spectrum of this derivative showed that the enol double bond was conjugated with the benzene ring, and the n.m.r. spectrum (only one hydrogen, H-4, at low field) ruled out the alternative structure (II), which would have had two hydrogens at low field, H-4 and H-12a (colchicine numbering).

Thus van Tamelen was able, at the expense of much exploratory work, to introduce the three-carbon unit without using the platform of the C-1 hydroxy group, which Eschenmoser used. The Eschenmoser route obviously has the advantage of elegance at this stage.

The next step was to introduce the remaining two carbon atoms of the carbon skeleton. A Reformatsky reaction gave the cyano ester (**5**) as a separable mixture of diastereoisomers. Saponification of each isomer gave the corresponding dicarboxylic acids (**6** and **7**).

The formation of the seven-membered ring was to be the next step, and an acyloin reaction was the method envisaged. But on just what derivative of **6** or **7** should this reaction be carried out? If the tertiary hydroxyl group were to be removed (for example, by dehydration and hydrogenation), the acyloin reaction would give a saturated ring, that is a ring at a low oxidation level. Oxidation up to the tropolone level was not an attractive proposition because of the ease with which the electron rich aromatic ring would be attacked by such likely oxidising agents as bromine. It was therefore going to be very helpful to keep to the oxidation level of **6** or **7**. Another possibility would be to use a dehydration product (III). The trouble with this derivative was that the most likely configuration at the double bond would be the wrong one, with the methoxycarbonyl group *trans* to the propionic ester group. Furthermore, the diradical intermediate (IV) of the acyloin

(III)　　　　　　　　　　　　　　　　　　(IV)

reaction had the possibility (IV, arrows) of forming a five-membered ring, five-membered rings being usually easier to get than their seven-membered counterparts.

What about the possibility of leaving the hydroxyl group as it was? This was sure to be troublesome, because it would be converted into its anion (V), which would immediately provide a strong and intramolecular base to catalyse a Dieckmann cyclisation (V → VI):

(V)　　　　　　　　　　　　　　　　　　　　　　　　(VI)

The derivative chosen was the lactone ester (**8** or **9**), in which the hydroxyl group was protected intramolecularly. In the course of an acyloin reaction, base is

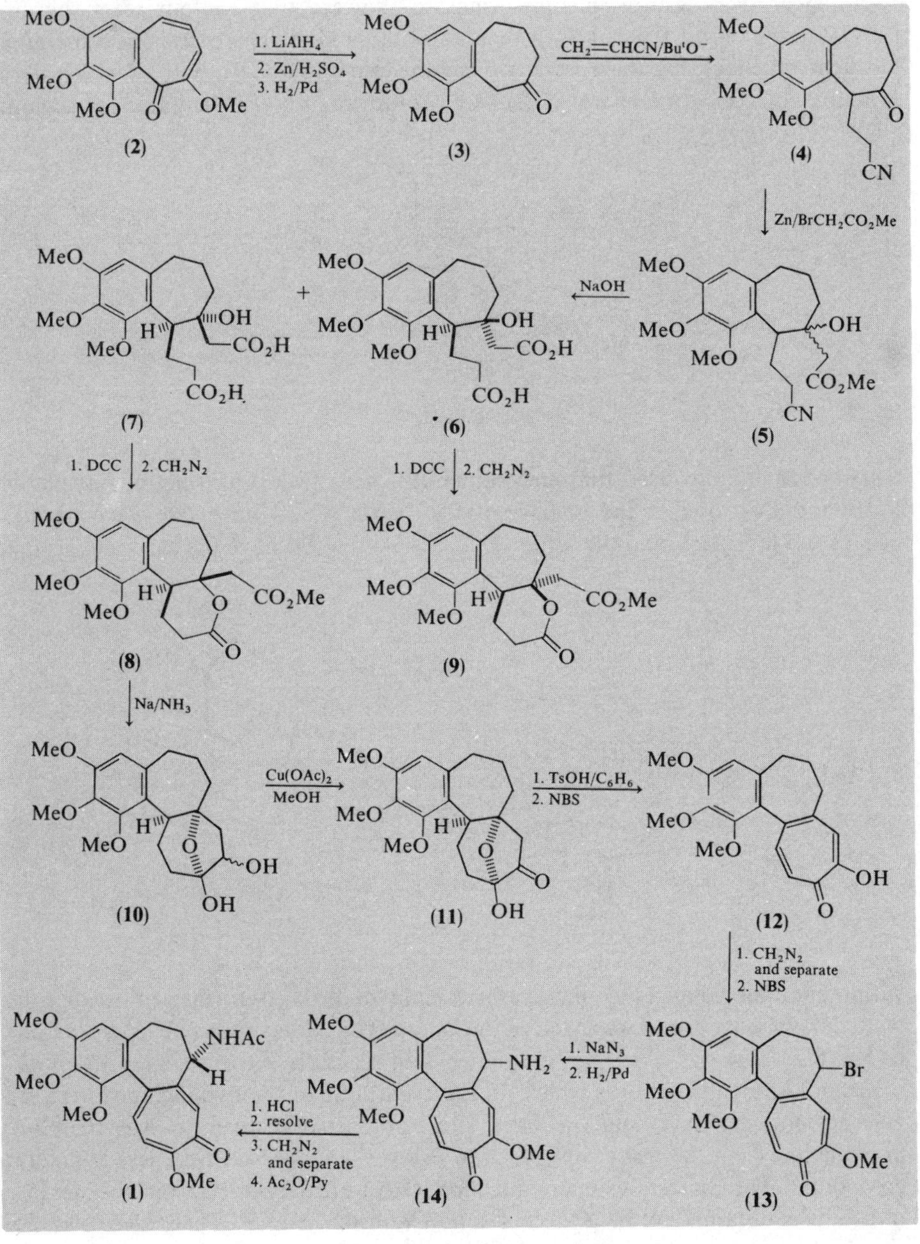
MeO
OMe
(2)
1. $LiAlH_4$
2. Zn/H_2SO_4
3. H_2/Pd
(3)
$CH_2{=}CHCN/Bu^tO^-$
(4)
CN
$Zn/BrCH_2CO_2Me$
(5)
OH
CO_2Me
NaOH
(6)
(7)
CO_2H
H
1. DCC 2. CH_2N_2
(8)
(9)
Na/NH_3
(10)
$Cu(OAc)_2$
MeOH
(11)
1. $TsOH/C_6H_6$
2. NBS
(12)
1. CH_2N_2 and separate
2. NBS
(13)
Br
1. NaN_3
2. H_2/Pd
(14)
NH_2
1. HCl
2. resolve
3. CH_2N_2 and separate
4. Ac_2O/Py
(1)
NHAc

created, which could cause the elimination of the acyloxy group β to the ester group. Fortunately there was ample precedent to indicate that acyloin reactions are usually faster than base-catalysed reactions of this kind. There was one remaining problem—the fact that there were two lactone esters (**8** and **9**), both of which were available, but only one of which (**8**) had a conformation (VII) suitable for acyloin

(VII)

reaction. In the *cis*-fused diastereoisomer (**9**), with two conformations available (VIII and IX), one of the conformations (VIII) would have the two carbonyl groups too far apart, and the other (IX), although it did have the carbonyl groups

(VIII)   (IX)

within bonding distance, would be a most unfavourable conformation because the trimethoxyphenyl group was axial in the lactone ring. Thus only one of the lactone esters (**8** $\equiv$ VII) was likely to give the acyloin reaction. At this stage, of course, it was not known which was which. In the event, one of the two lactone esters did give acyloin products, and the other did not. Retrospectively it was therefore possible to assign the *trans* configuration (**8**) to the higher-melting, less abundant compound. The diastereoisomeric alcohols (**10**) were separated from the mixture of products obtained in the acyloin reaction with this isomer. These alcohols were simply the hemiacetals of the expected α-hydroxyketones.

Oxidation of this mixture gave the ketone (**11**), which was dehydrated with acid and oxidised to desacetamidocolchiceine (**12**).

This sequence used by van Tamelen for making the tropolone ring had the advantage that the correct disposition of oxygen atoms was obtained directly. The remaining steps resembled those used by Eschenmoser, except that azide ion was used as the nitrogen nucleophile, and the intermediate azide was converted to the amine (**14**) by catalytic hydrogenation.

References

1. E. E. van Tamelen, T. A. Spencer, D. S. Allen and R. L. Orvis, *Tetrahedron*, **14**, 8 (1961).

Colchicine (Third Synthesis)

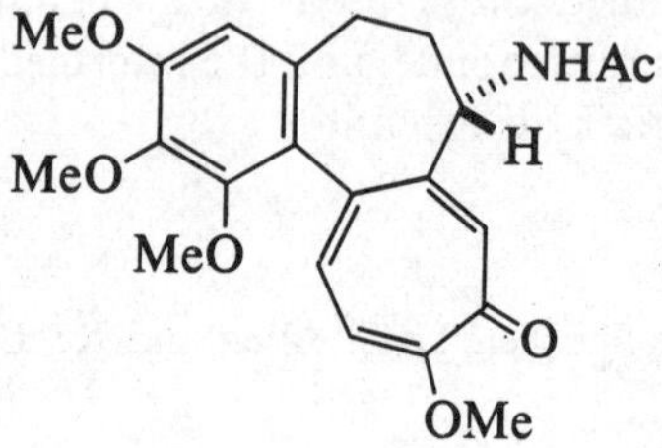

(1)

Another synthesis, by Scott,[1] also began with purpurogallin (2), and also took advantage of the conversion of desacetamidocolchiceine to colchicine which had been established in the earlier syntheses. But there the resemblance ends, because Scott used the tropolone ring of purpurogallin to provide ring C.

Scott's route was based on a proposal, one of several, for the biosynthesis of colchicine. The proposal was that a phenolic coupling (e.g. I) occurs to join the benzene ring to the tropolone ring. This mechanism is an application of the fundamental ideas of Barton[2] on phenolic coupling and its role in biosynthesis. As it

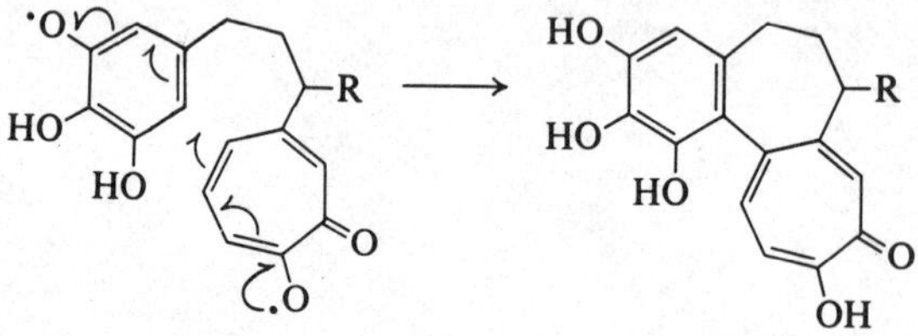

(I)   (II)

happens, it is now known that a phenolic coupling is involved in the biosynthesis of colchicine; but it occurs before the expansion of ring C.[3] Nevertheless, Scott's synthesis of colchicine, although it does not follow the biosynthetic route, is still biosynthetically patterned (see also pp. 23 and 156).

Purpurogallin (2) was known to be oxidised to the dicarboxylic acid (3), which could be dehydrated to the anhydride (4). It was also known that this enol derivative would condense with aldehydes, but the reaction with 3,4,5-trimethoxyphenyl-acetaldehyde (8) proved refractory. It was eventually found that simply heating a mixture of the reagents gave the lactone (9). In the course of this reaction decarboxylation occurred, as a result of the influence of the carbonyl group in the tropolone ring (III → IV):

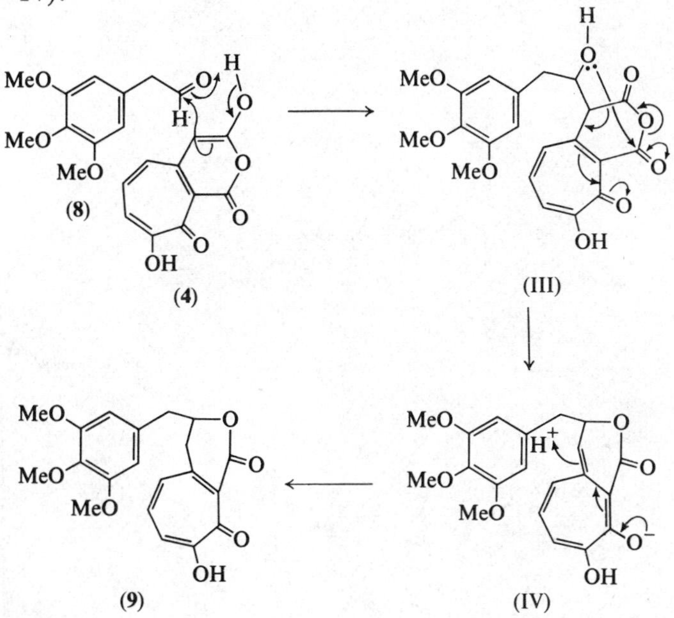

Further pyrolysis of the lactone (9) caused elimination and decarboxylation to give the styrene (10). Hydrogenation and demethylation gave the pyrogallol (11). This compound was then ready for oxidation—in the sense (I) which had been the reason for trying this approach. The oxidation of so sensitive a compound as a pyrogallol had not been the first thought; it was tried only after much earlier experimentation with compounds having fewer free hydroxyl groups. It was expected that it might be quite difficult to preserve the immediate product (II, R = H, 12) under the conditions of the oxidation; so, before oxidation of 11 was attempted, the expected product was prepared by demethylation of desacetamidocolchicine. Since remethylation gave 13, and since 13 had already been converted to colchicine in the earlier syntheses, the only reaction needed to complete the synthesis was the oxidation of 11 to 12.

The sensitivity of 12 to oxidation proved to have been indeed a cause for concern. With most oxidising agents the desired product (12) was too rapidly decomposed to offer much hope of isolating it. But with potassium ferricyanide and with ferric chloride the decomposition of 12 was not excessively fast or disruptive. With the former reagent, the product was V, which was not helpful in itself because the aromatic ring had been broken, no doubt irreversibly. But it was encouraging

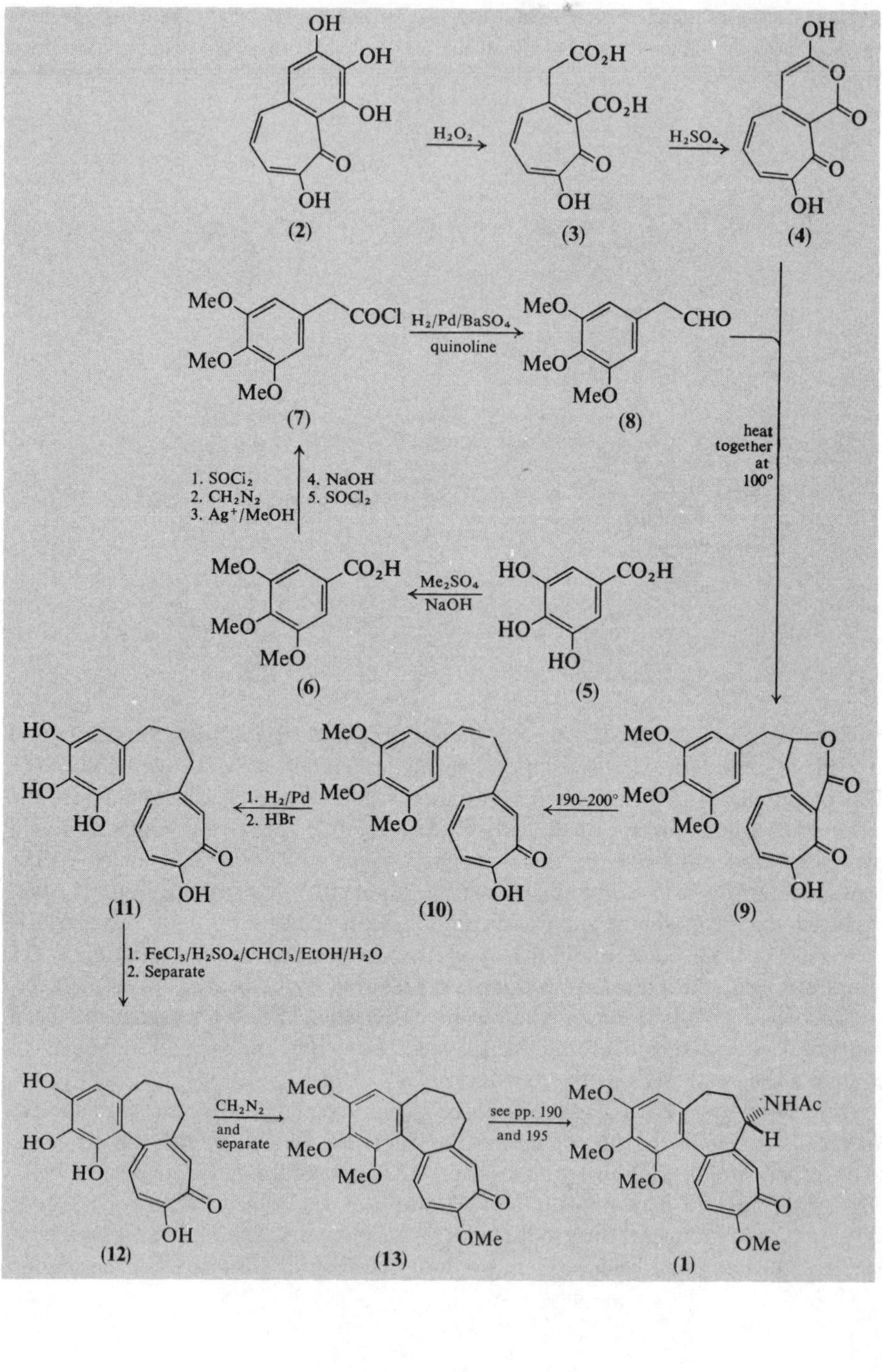

(2) $\xrightarrow{H_2O_2}$ (3) $\xrightarrow{H_2SO_4}$ (4)

(7) $\xrightarrow{H_2/Pd/BaSO_4 \atop quinoline}$ (8)

1. SOCl₂   4. NaOH
2. CH₂N₂   5. SOCl₂
3. Ag⁺/MeOH

(6) $\xleftarrow{Me_2SO_4 \atop NaOH}$ (5)

heat together at 100°

(11) $\xleftarrow{1.\ H_2/Pd \atop 2.\ HBr}$ (10) $\xleftarrow{190-200°}$ (9)

1. FeCl₃/H₂SO₄/CHCl₃/EtOH/H₂O
2. Separate

(12) $\xrightarrow{CH_2N_2 \atop and \atop separate}$ (13) $\xrightarrow{see\ pp.\ 190 \atop and\ 195}$ (1)

(12)   (V)   (11)

to find that the same product (V) was obtained from **11** when it was treated with potassium ferricyanide. This observation showed that the *idea* of oxidative coupling was feasible. Unfortunately, the oxidation of **11** was about a thousand times slower than that of **12**, so there was no hope of isolating the latter from an oxidation of the former using this oxidising agent. Eventually conditions were found for the conversion of **11** to **12**. Scott used ferric chloride and a two-phase mixture of chloroform and acidic aqueous ethanol for the oxidation, and paper chromatography in a nitrogen atmosphere for the isolation of the product (**12**). It is quite likely that the oxidative coupling step is not a diradical coupling (I → II), because it is known that tropolones enter into radical coupling reactions only under comparatively vigorous oxidative conditions. An ionic process (VI arrows) seems a likely candidate for the carbon–carbon bond forming step. The yield was naturally low, 4–5 per cent; but a total synthesis had been completed.

(VI)

Question

You will have noticed that in the first two syntheses the *reduction* of purpurogallin derivatives took place in the *tropolone* ring, whereas the *oxidation* of purpurogallin in this synthesis took place in the *benzene* ring. While the methyl groups present in the first two syntheses will have affected the position somewhat, this trend is what would have been expected. What are the features of the two rings which lead them to show this selectivity?

References

1. A. I. Scott, F. McCapra, R. L. Buchanan, A. C. Day and D. W. Young, *Tetrahedron*, **21**, 3605 (1965); see also, A. I. Scott, F. McCapra, J. Nabney, D. W. Young, A. C. Day, A. J. Baker and T. A. Davidson, *J. Amer. Chem. Soc.*, **85**, 3040 (1963).
2. D. H. R. Barton and T. Cohen, *Festschrift A. Stoll*, Birkhauser, Basle, 1957, p. 117.
3. A. R. Battersby, *Pure Appl. Chem.*, **14**, 117 (1967).

Colchicine (Fourth Synthesis)

(1)

The fourth synthesis of colchicine is due to Woodward,[1] and is entirely different from the other three. It does not use purpurogallin, and it does not rely on the bromination-ammonolysis reaction to insert the nitrogen atom. Instead, the nitrogen atom is present from the start in the form of an isothiazole ring. When this work was begun, isothiazoles were unknown. They have received some attention since, but at least one of the fascinations of this synthesis is the revelation it provides of some of the chemistry of these compounds.

The route used to prepare the isothiazole (**3**) was itself quite remarkable. When β-aminocrotonate (**2**), prepared from methyl acetoacetate and ammonia, was treated with thiophosgene, the isothiazole (**3**) was obtained directly. The expected product (I) had presumably been formed and had then cyclised by a mechanism

involving a higher oxidation state of sulphur. It thus achieved aromaticity at the expense of forming a zwitterion (III) bearing a formal negative charge on C-3. The ease of this reaction implied that an intermediate (such as III), bearing a negative charge on C-3, was a relatively stabilised one. That a negative charge at this site should be easily got was not too surprising, since it was well known that the corresponding anion in thiazole chemistry (e.g IV obtained from thiazoles, and V obtained from thiazolium salts) was a stabilised system. (This stabilisation—in the zwitterion V—is important in explaining the mechanism of thiamine action.)

(IV)　　　　(V)

If such anions were indeed available in the isothiazole series, as the ease of formation of III would seem to suggest, then it should be possible later in the synthesis to generate an anion at C-3 of the isothiazole and use it as a nucleophile. This was to be an important part of the overall plan; but first a series of transformations on the substituents already on the isothiazole ring was carried out.

N-Bromosuccinimide caused bromination on the methyl group of 3, and triphenylphosphine then gave the phosphonium salt (4). Wittig reaction using the ylid of this salt and 3,4,5-trimethoxybenzaldehyde gave the olefin (6). It was then necessary to hydrogenate the double bond; but, because this was a sulphur-containing compound, the usual catalytic method was not applicable. Fortunately, diimide was well suited for this situation. Diimide (HN=NH) was readily available (as a reactive intermediate, that is) by several routes; the method used here was the copper-catalysed oxidation of hydrazine. Diimide, generated *in situ* by this method, reduced the double bond to give the ester (7). The ester group was reduced and reoxidised to give the corresponding aldehyde (8). This aldehyde was used in a Wittig reaction with the ylid (9). Ester hydrolysis and iodine-catalysed isomerisation of the double bonds gave the all *trans* acid (10). The next step established ring B, the first of the seven-membered rings: acid-catalysed cyclisation gave the β,γ-unsaturated acid (11), as a mixture of isomers about the double bond. The carbon–carbon bond forming step (VI) was of the Friedel-Crafts type, and the intermediate enol (VII) was no doubt protonated at the kinetically favoured site (c.f. p. 161).

(VI)　　　　(VII)　　　　(11)

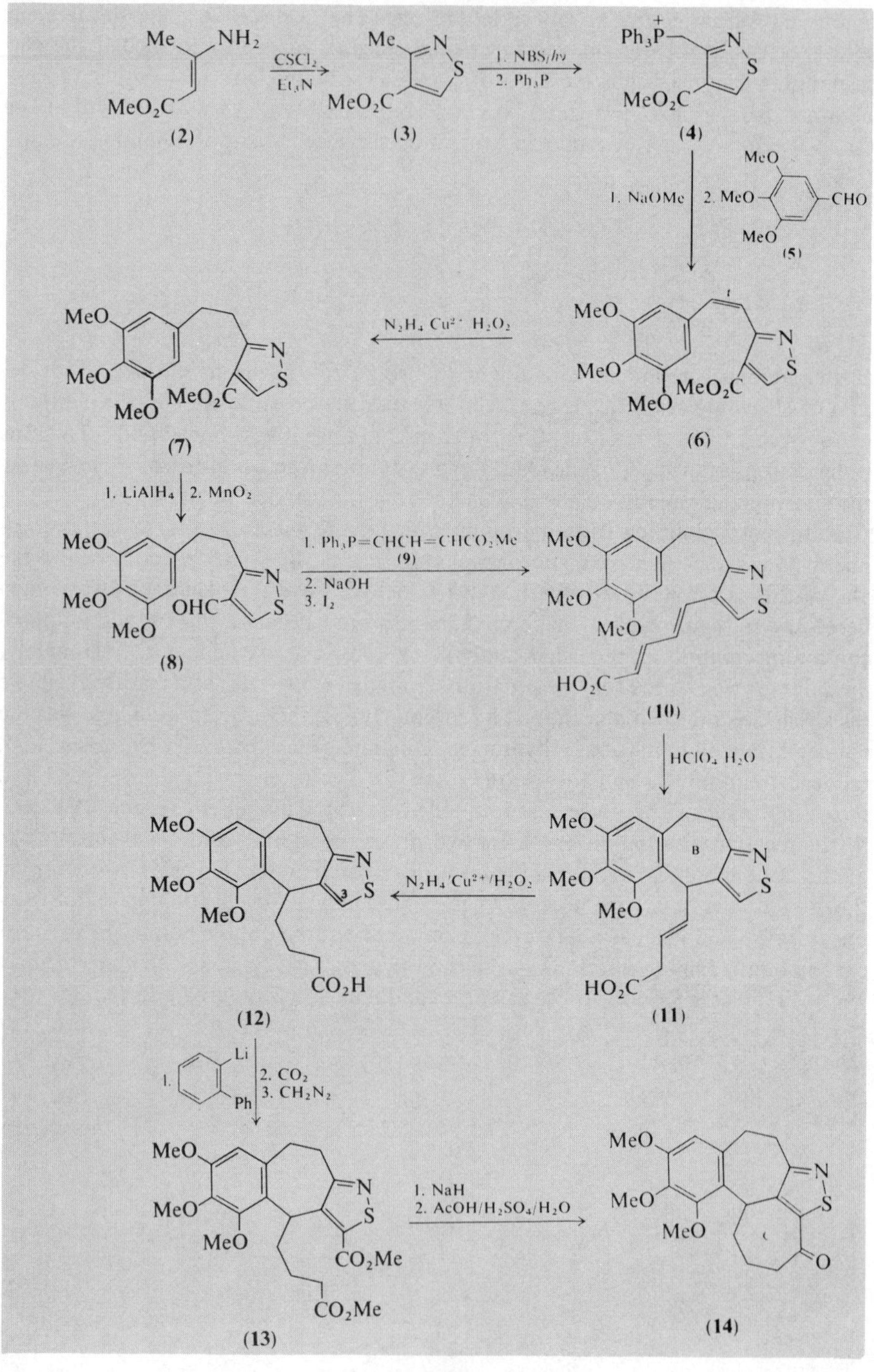
Me
NH2
MeO2C
(2)
CSCl2
Et3N
(3)
1. NBS/hν
2. Ph3P
Ph3P
(4)
1. NaOMe
2. MeO
CHO
(5)
(6)
N2H4 Cu2+ H2O2
(7)
1. LiAlH4
2. MnO2
OHC
(8)
1. Ph3P=CHCH=CHCO2Me
(9)
2. NaOH
3. I2
HO2C
(10)
HClO4 H2O
B
(11)
N2H4/Cu2+/H2O2
CO2H
(12)
Li
Ph
2. CO2
3. CH2N2
CO2Me
(13)
1. NaH
2. AcOH/H2SO4/H2O
C
(14)

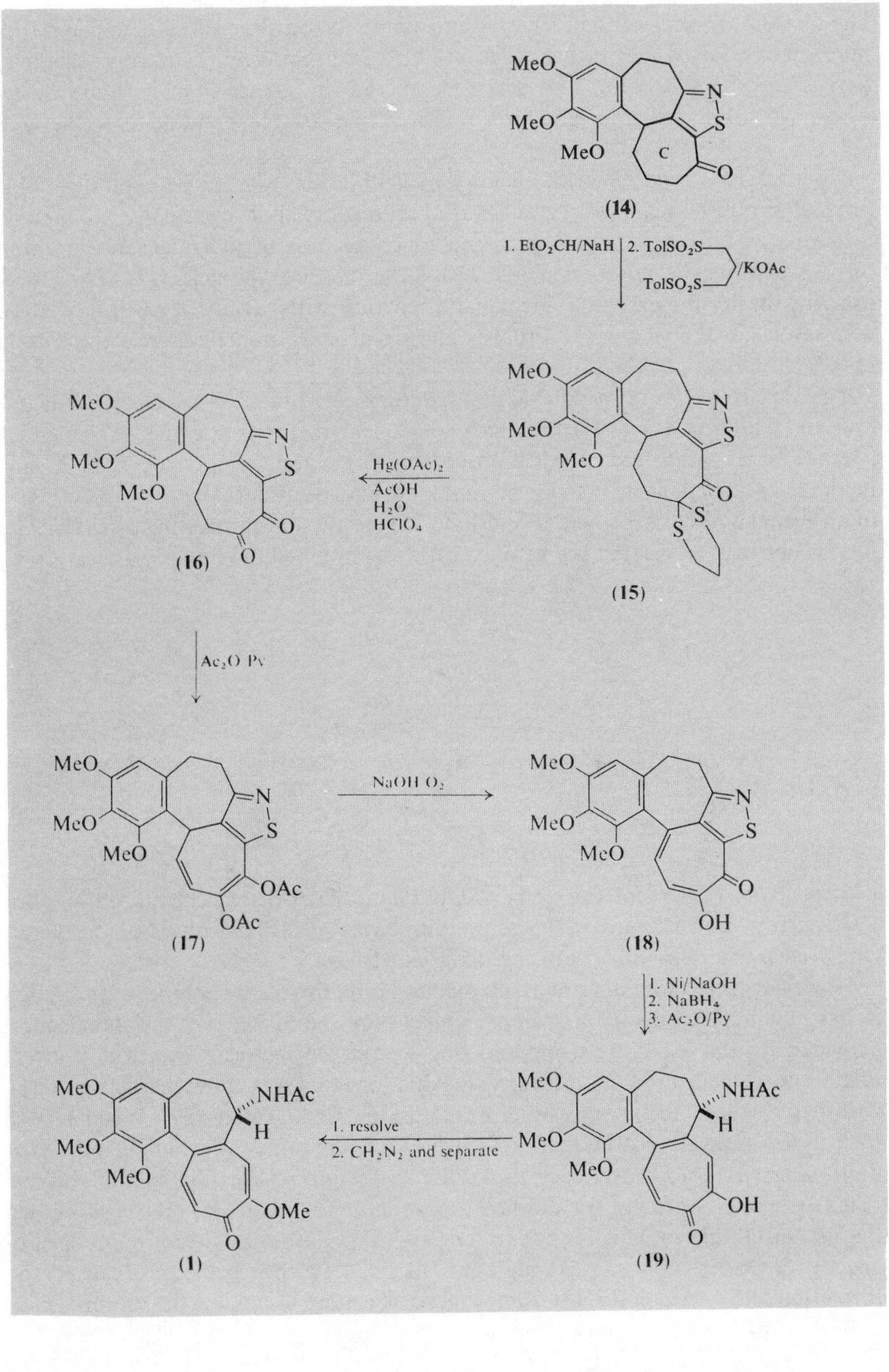
MeO
MeO
MeO
C
N
S
O
(14)
1. EtO2CH/NaH
2. TolSO2S
TolSO2S
/KOAc
(15)
Hg(OAc)2
AcOH
H2O
HClO4
(16)
Ac2O Py
(17)
OAc
NaOH O2
(18)
OH
1. Ni/NaOH
2. NaBH4
3. Ac2O/Py
(19)
NHAc
H
1. resolve
2. CH2N2 and separate
OMe
(1)

The fact that several of the carbon atoms going to form the seven membered ring were sp^2 hybridised meant that they were held in a relatively rigid array: this may have made the formation of a seven-membered ring easier than it would otherwise have been. Diimide reduction removed the double bond to give the acid (**12**).

Now was the time when the acidity of isothiazoles was to be exploited. This proved troublesome, because of variable results; but eventually *o*-biphenyl lithium was found to be suitable, and carboxylation of the intermediate anion on C-3 (isothiazole numbering) then gave the dicarboxylic acid (**13**). The correponding diester underwent a Dieckmann reaction with, again, smooth formation of a seven-membered ring. Hydrolysis and decarboxylation then gave the ketone (**14**).

The carbon skeleton of colchicine was complete, but ring C was at much too low an oxidation level. Furthermore, the usual methods for the oxidation of ketones, such as the use of selenium dioxide, were not notably selective. A new method for achieving selectivity of this kind had been developed by Woodward in an earlier synthesis,[2] where it had functioned only in a protecting capacity. The formyl derivative (VIII) of the ketone (**14**) was combined, with mild base catalysis,

(VIII)   (IX)

with propane-1,3-dithiol di-*p*-toluenesulphonate. Two displacements on sulphur (VIII, arrows and IX, arrows) gave the dithioketal (**15**). Hydrolysis of this ketal, catalysed by mercuric ion, gave the diketone (**16**).

However, ring C was still not easily oxidised directly to a tropolone, even though it was now only a dihydro tropolone. This was probably because the stabilisation provided by the aromatic tropolone ring would only appear late in any direct oxidation. To get round this difficulty, the diketone (**16**) was converted, by warming with pyridine and acetic anhydride, into a double enol acetate (**17**). When **17** was hydrolysed with alkali, the first-formed diol (X) contained a grouping—the enediol—which is notoriously easy to oxidise, especially under alkaline conditions. (This grouping is present for example, in vitamin C (see p. 36), the well-known thermal instability of which is due to the ease of oxidation by the air of an enediol group.) Simply passing air through the alkaline hydrolysis mixture caused the formation of the tropolone (**18**). Raney nickel was used to remove the sulphur, and, because the tropolone anion was more stable towards reduction than a free tropolone, it was used in alkaline solution. Reduction of the double bond to nitrogen was completed by the addition of sodium borohydride; and the product, isolated after acetylation, was colchiceine (**19**). The remaining steps to colchicine, including resolution, were already known.

Questions

1. How would you synthesise the ylid (**9**)?
2. The displacement of *p*-toluenesulphinate is not as common as the displacement of *p*-toluenesulphonate; nevertheless there is ample precedent for the displacement of *p*-toluenesulphinate, which occurs in the step (**14 → 15**). Do you know of any precedent?

References

1. R. B. Woodward (with G. Volpp and J. Z. Gougoutas), *The Harvey Lectures*, 31 (1963).
2. R. B. Woodward, A. A. Patchett, D. H. R. Barton, D. A. J. Ives and R. B. Kelly, *J. Chem. Soc.*, 1131 (1957).

Prostaglandins F$_{2\alpha}$ (1) and E$_2$ (2)

(1)   (2)

The prostaglandins are a group of hormones with a wide range of activity and great potential value in medicine.[1] They are present in most mammalian tissues in very low concentration, the highest concentration being in human semen. In view of their potential importance in the control of such things as blood pressure, uterine contractions, and fertility, and in view of the very low concentrations available from natural sources, considerable impetus has been given to the prospect of obtaining these compounds by total synthesis. Although they have complex structures and require many-stage syntheses, the quantities needed are comparatively small. Synthesis is therefore a very attractive proposition as a way of obtaining pure materials. A major problem, however, is that the prostaglandins are unstable in acid or base. For example, the β-hydroxyketone group in 2 is very easily dehydrated—only the very mildest hydrolytic conditions can be used while this group, or the 1,3-diol group of 1, is present. Nevertheless, many syntheses have been reported, several of these being by E. J. Corey and his group at Harvard. The most versatile and the most beautiful is this one,[2] in which great stereochemical control was achieved.

The anion of cyclopentadiene was alkylated at −55° to give the methoxymethylcyclopentadiene (3). The diene was combined with 2-chloroacrylonitrile in a Diels-Alder reaction, using cupric fluoroborate as a catalyst, to give the adduct (4). We should note two things about this reaction. Firstly, the use of a catalyst in pericyclic reactions is comparatively rare; such reactions, with cyclic transition states, do not usually offer much opportunity for acids or bases to intervene and, in so doing, to speed up the reaction. However, it is well known that Diels-Alder reactions of electron deficient olefins, such as maleic anhydride, go faster than do the reactions of unsubstituted olefins. It is likely that what the Lewis acid catalysis does (I) is to make the nitrile group more electron withdrawing than it already is. Secondly, this catalysis is necessary because the Diels-Alder reaction

(I)

must be made to go faster than the other reaction which cyclopentadienes, such as **3**, readily undergo. This unwanted reaction is the rearrangement which gives the isomers (III and IV):

(II)   (III)   (IV)

This rearrangement is also a pericyclic reaction, in fact it is a 1,5 suprafacial sigmatropic shift. And it is a reaction which goes quite rapidly in cyclopentadienes at room temperature.

By using low temperatures, and by using the catalyst to accelerate the Diels-Alder reaction (but not the sigmatropic hydrogen shift), it was possible to prepare and use the diene (**3**) to give the adduct (**4**), in over 90 per cent yield,[3] as a mixture of stereoisomers differing only in the orientation of the chloro and cyano groups. The methoxymethyl group was expected, on steric grounds, to be on the side of the five-membered ring opposite to that attacked in the Diels-Alder reaction. That this had indeed been the case was proved by the subsequent transformations into natural prostaglandins.

Alkali treatment converted the mixture (**4**) to a single ketone (**5**), which was treated with peracid to give the Baeyer-Villiger product (**6**), without concomitant epoxidation of the double bond (which was comparatively hindered). In the Baeyer-Villiger reaction, an intermediate (V) is formed, in which alkyl migration

(V)

to electron deficient oxygen takes place in a reaction analogous to the Beckmann rearrangement. In reactions of this kind, it is usual to find that the group migrating is the one which is better able to support a partial positive charge. This follows from

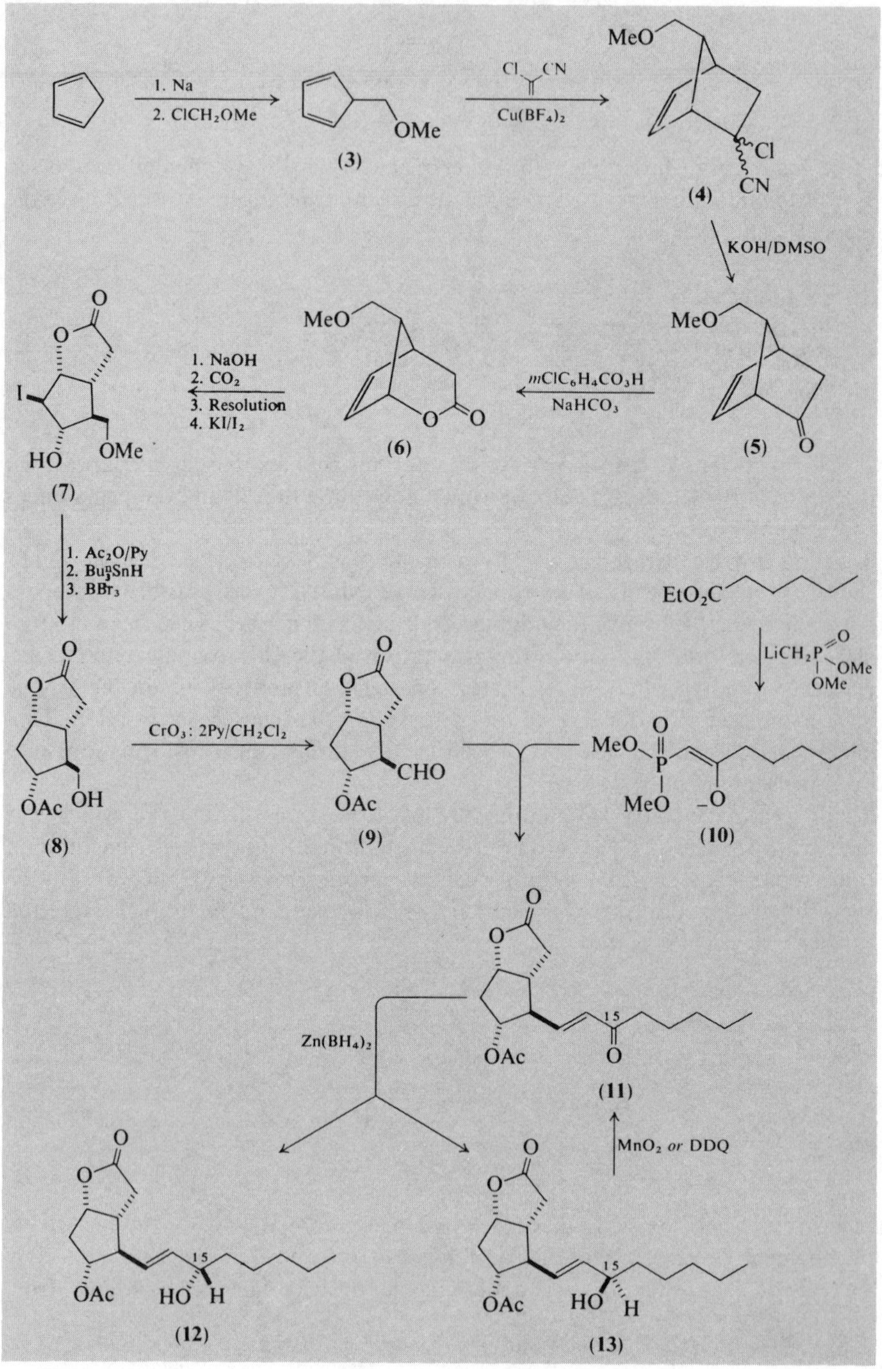
1. Na
2. ClCH2OMe
OMe
(3)
Cl CN
Cu(BF4)2
MeO
Cl
CN
(4)
KOH/DMSO
MeO
O
(5)
mClC6H4CO3H
NaHCO3
MeO
O
O
(6)
1. NaOH
2. CO2
3. Resolution
4. KI/I2
O
O
I
HO
OMe
(7)
1. Ac2O/Py
2. Bu3nSnH
3. BBr3
EtO2C
LiCH2P O OMe OMe
O
O
OAc
OH
(8)
CrO3: 2Py/CH2Cl2
O
O
CHO
OAc
(9)
MeO
P
MeO
O
(10)
O
O
15
OAc
O
(11)
Zn(BH4)2
MnO2 or DDQ
O
O
15
OAc
HO
H
(12)
O
O
15
OAc
HO
H
(13)

the nature of the transition state, in which the positive charge is partially spread onto the carbon atom undergoing migration.

$$R \diagdown \overset{+}{X} \longrightarrow R \diagdown \underset{\diagup}{\overset{\diagdown}{\underset{+X}{}}} \longrightarrow \underset{+}{\diagup} \overset{R}{\underset{\diagdown X}{|}}$$

Thus in the case of **5** it is the secondary alkyl group which migrates, not the primary one.

Saponification of the lactone (**6**) gave a hydroxy acid, which could easily be resolved. Treatment of the salt of the acid with potassium triiodide gave the iodolactone (**7**). The intermediate hydroxy acid (VI) participated in the iodination step, with the result that a single isomer was produced, having five chiral centres, one at each carbon atom of the five-membered ring.

Acetylation and reduction to remove the iodine atom, and boron tribromide reaction to remove the methyl ether, gave the alcohol (**8**). Oxidation gave the aldehyde (**9**), which was combined with the anion of dimethyl 2-oxoheptyl phosphonate (**10**) to give the α,β-unsaturated ketone (**11**). This variation of the Wittig reaction was introduced by Wadsworth and Emmons,[5] and is known to give the *trans* double bond. Reduction of the ketone with zinc borohydride gave a mixture of the 15α and 15β alcohols (**12** and **13**), which were separated chromatographically. The 15β isomer (**13**) could be oxidised back to the ketone (**10**) and was therefore not wasted. This is the only step where no control over stereochemistry was exercised, or could reasonably be expected to have been exercised.[6]

The acetate protecting group was removed and both hydroxyl groups protected as tetrahydropyranyl ethers (**14**). Partial and careful reduction of **14**, using diisobutyl aluminium hydride, gave the hemiacetal (**15**). The capacity of this reducing agent to convert lactones to hemiacetals had been discovered earlier by Wettstein.[8] A Wittig reaction of the (masked) aldehyde group in **15** with the ylid (**16**) of 5-triphenylphosphoniopentanoic acid gave the olefinic acid (**17**) with a *cis* double bond. The formation of a *cis* double bond had been expected, by analogy with

Corey's earlier work using dimethylsulphoxide as a solvent for the Wittig reaction. Removal of the protecting group gave prostaglandin $F_{2\alpha}$ (1).

Alternatively, oxidation of **17**, followed by removal of the protecting groups, gave prostaglandin E_2 (**2**). Further versatility in this synthesis is demonstrated by the fact that the *cis* double bond of the bis(isopropyldimethylsilyl) ether of **2** could be hydrogenated selectively to give prostaglandin E_1 (the subscript 1 implying that the acidic side chain is saturated).[9]

It is amusing to note that the apparently simpler prostaglandins with the saturated acidic side chain, should turn out to be more easily synthesised by this route than by earlier routes which made no attempt to tackle the problem presented by the extra double bond.

Questions

1. The general rule, for reactions such as the Baeyer-Villiger and Beckmann rearrangements, that the group better able to support positive charge migrates more readily, is often broken when one of the groups is vinyl or phenyl. For example, acetophenone gives phenyl acetate and not methyl benzoate. Nevertheless, of the two oxime tosylates (VII and VIII), only the latter rearranges easily (at room

(VII)   (VIII)

temperature in fact) whereas the former can be distilled at $200°/0.3$ mm, unchanged. What feature, which is not available to VII, enables the phenyl group to migrate in the oxidation of acetophenone? For this question, the difference between a phenyl group and the vinyl group of VII is not important: the main point is that phenyl and vinyl are both high-energy cations and should not, by the usual rule, migrate with ease. In the one case phenyl migrates by preference but in the other the vinyl group is strikingly loth to migrate.

2. How would you synthesise the ketone which was used to prepare the oxime tosylates (VII and VIII)?

References

1. S. Bergström, *Science*, **157**, 382 (1967).
2. E. J. Corey, N. M. Weinshenker, T. K. Schaaf and W. Huber, *J. Amer. Chem. Soc.*, **91**, 5675 (1969).
3. Corey[4] has subsequently improved this synthesis by using the thallium salt of cyclopentadiene in place of the sodium salt. This was particularly helpful for the large scale operations now being carried out in several industrial laboratories.
4. E. J. Corey, U. Koelliker and J. Neuffer, *J. Amer. Chem. Soc.*, **93**, 1489 (1971).
5. W. S. Wadsworth and W. D. Emmons, *J. Amer. Chem. Soc.*, **83**, 1733 (1961).
6. These incautious words were written only shortly before a paper was published by Corey[7] showing that a specially designed reducing mixture, comprising diisopinocamphenyl-t-butylborohydride in ether-THF containing hexamethylphosphoramide, was capable of giving, on a slightly different derivative of **11**, 68% of the 15α alcohol, and 31% of the 15β alcohol. More recently still, Corey, K. B. Becker and R. K. Varma, *J. Amer. Chem. Soc.*, **94**, 8616 (1972) have improved this ratio to 92:8 by using a rationally chosen *p*-phenylphenylurethane group in place of the acetyl group in **11**.
7. E. J. Corey, S. M. Albonico, U. Koelliker, T. K. Schaaf and R. K. Varma, *J. Amer. Chem. Soc.*, **93**, 1491 (1971).
8. J. Schmidlin and A. Wettstein, *Helv. Chim. Acta*, **46**, 2799 (1963).
9. E. J. Corey and R. K. Varma, *J. Amer. Chem. Soc.*, **93**, 7319 (1971).

Reserpine

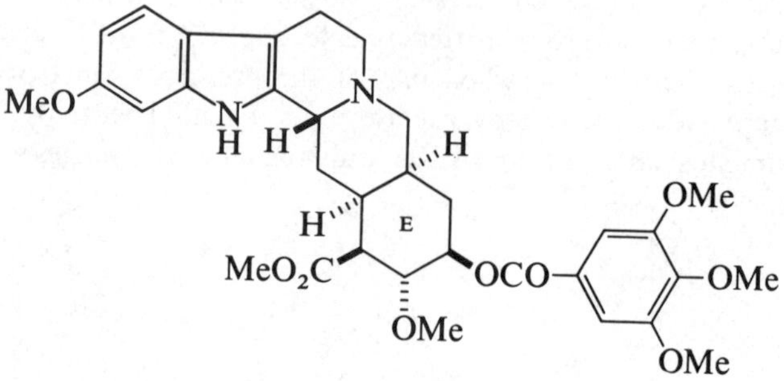

(1)

The penultimate chapter of this book has been devoted to a particularly striking Corey synthesis. There is no doubt that the final chapter must belong to Woodward. His extraordinary syntheses, such as those of quinine,[1] strychnine,[2] reserpine,[3] lysergic acid,[4] tetracycline,[5] cephalosporin C[6] and chlorophyll,[7] present so many facets of interesting chemistry that a book could easily have been written about them alone. I have chosen for the final chapter the synthesis[3] of the alkaloid reserpine because it presents a particularly beautiful solution to the problem which is automatically created by the presence in that molecule of six chiral centres, a problem in synthesis which I have stressed throughout this book and which here finds its most accomplished solution.

Reserpine is an alkaloid which has found much use in the treatment of hypertensive disorders. The chiral centres are concentrated in ring E, where five of the six reside. The first target was therefore an intermediate possessing the features (I) which were needed for ring E and could be extended (at X and Y) to give reserpine itself.

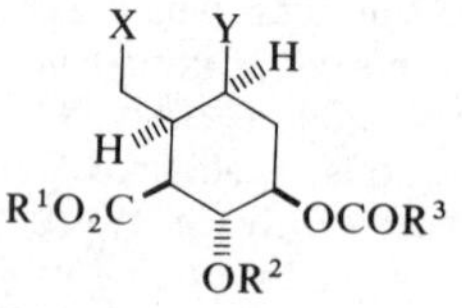

(I)

Diels-Alder reaction between the diene **(2)** (prepared from acrolein and malonic acid followed by esterification) and quinone gave in one step an adduct **(3)** with three chiral centres correctly set up. This was a consequence of the well known preference for *endo* addition (II) in Diels-Alder reactions, a preference only recently accounted for in the Woodward-Hoffmann theories for concerted reactions.

(II)   (III)

Reduction of the diketone (3) with aluminium isopropoxide in isopropanol gave the lactone (4), readily distinguished (by its i.r. spectrum) as the 5-membered ring lactone and not the 6-membered ring lactone (which would have been possible if the other hydroxyl group had displaced methanol from the ester group). The two fused six-membered rings of 3 folded (III), as they do in *cis*-decalins, for example, to make the inner, or concave, surface relatively hindered. Hence the isopropoxide delivered hydride to the less hindered side of the carbonyl groups: the convex surface. There were now, in 4, two isolated double bonds. Preliminary experiments had indicated that these would be of different reactivity towards electrophiles. The lower double bond was found to react with bromine to give the bromoether (5). (It is now well known that a suitably placed hydroxy group can participate in attack by bromine in the sense shown by IV.) The presence of the

(IV)   (V)   (VI)

β-bromocarbonyl system in 5 made base-catalysed elimination of HBr easy (V; X = Br; arrows); the intermediate unsaturated lactone (VI) then underwent addition of methoxide ion to give the ether (6).

The addition of methoxide could reliably be expected to take place on the convex surface of VI, especially since the concave surface was now encumbered with two bridges. It was not quite so clearly predictable that protonation of the enolate (V; X = OMe) would also take place on the convex surface, but models suggested that this structure was probably the less strained one. Also, since this centre could be equilibrated under the conditions of the reaction, and only one isomer was found amongst the products, it seemed very likely that this was indeed the isomer in hand.

The molecule thus prepared contained the five contiguous chiral centres of ring E, correctly disposed. This sequence, leading to a compound of the complexity of **6**, is extraordinarily short. It is short partly because of the elegance of the conception and partly because of the way in which information gained from earlier work with related routes was put to work to achieve the simplicity of the one given above. There are not many occasions in which only three stereoselective reactions are used to give a molecule with seven chiral centres. The remaining steps are involved with the elaboration of **6** towards the fragment (I) needed to join on to the tryptamine portion of the molecule.

N-Bromosuccinimide with sulphuric acid gave the bromohydrin (**7**). This step actually introduced, temporarily but stereospecifically, two more chiral centres. The electrophilic bromine necessarily attacked from the convex surface of **6**. The site of the attack of water was rather less predictable but a model of **6** shows that antiplanar attack, as in VII → VIII, would lead smoothly to the diaxial product,

(VII)   (VIII)

whereas if water were to attack at the other carbon atom a boat conformation would have to be produced in the first instance.

Oxidation of the alcohol to the ketone followed by metal reduction gave the unsaturated ketone, which was acetylated to give **9**. One of several possible pathways for this reduction is shown as IX → X → XI.

(IX)   (X)   (XI)

Osmium tetroxide gave the diol corresponding to **9**. Cleavage with periodic acid gave, presumably, the aldehydo acid (**10**). This was not isolated but was methylated to the corresponding diester, then condensed with 6-methoxytryptamine, and the

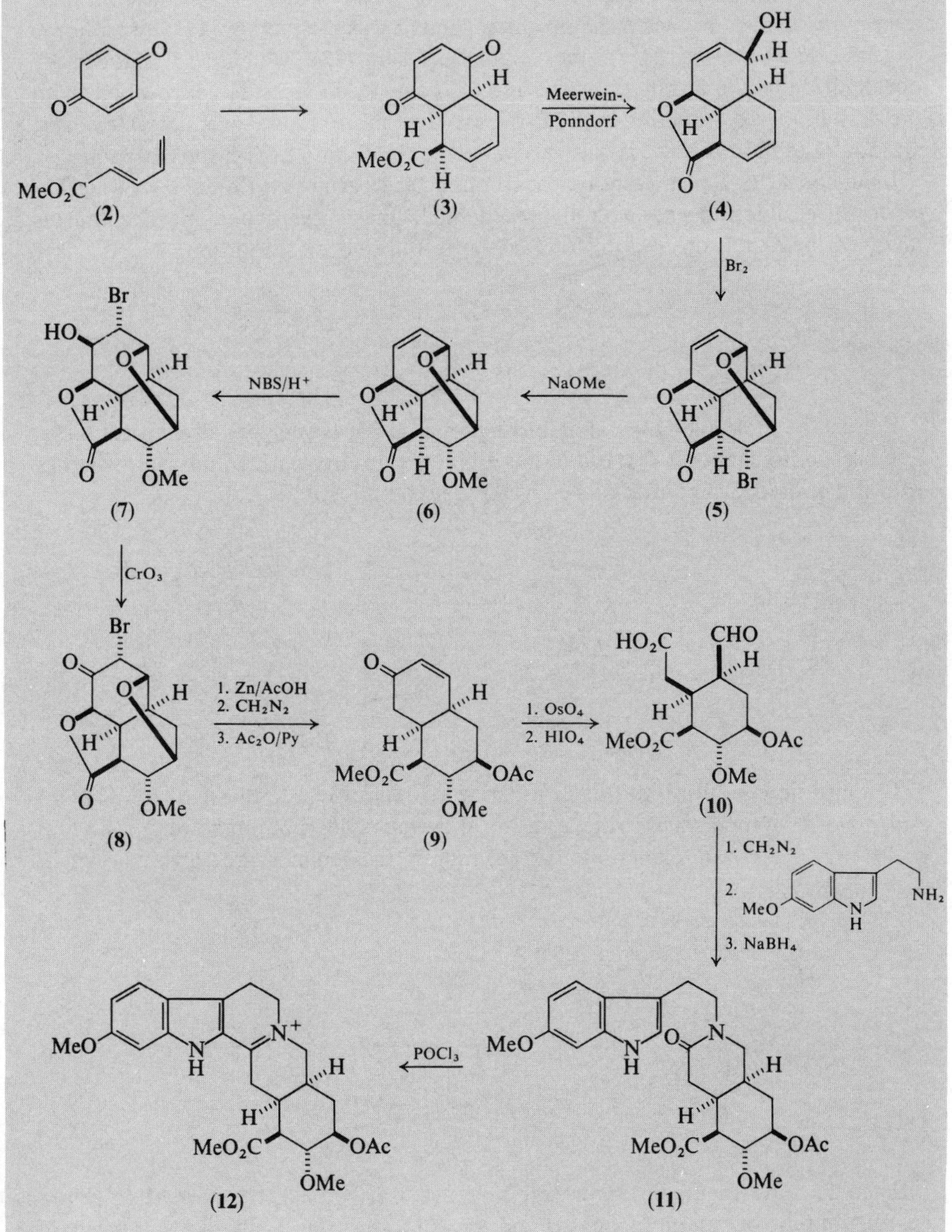

MeO2C
(2)
H
MeO2C
H
(3)
Meerwein-Ponndorf
OH
H
H
O
H
O
(4)
Br2
O
H
O
H
O
H
Br
(5)
NaOMe
O
H
O
H
O
H
OMe
(6)
NBS/H+
Br
HO
O
H
O
H
O
OMe
(7)
CrO3
Br
O
O
H
O
H
O
OMe
(8)
1. Zn/AcOH
2. CH2N2
3. Ac2O/Py
O
H
H
MeO2C
OAc
OMe
(9)
1. OsO4
2. HIO4
HO2C
CHO
H
H
MeO2C
OAc
OMe
(10)
1. CH2N2
2.
MeO
N
H
NH2
3. NaBH4
MeO
N
H
O
N
H
H
MeO2C
OAc
OMe
(11)
POCl3
MeO
N
H
N+
H
H
MeO2C
OAc
OMe
(12)

intermediate imine was reduced with sodium borohydride. The resulting amino ester cyclised instantly, and the product isolated was the lactam (**11**). Phosphorus oxychloride converted **11** to the quaternary salt (**12**), which was reduced by borohydride to the tetrahydrocarboline (**13**), in which hydride approached from the less hindered direction and also gave the more stable product. However, although the more stable, it was the wrong configuration at C-3 for reserpine.

Inversion at C-3 of a tetrahydrocarboline by treatment with acid was *mechanistically* feasible and was also well precedented. One of several possible mechanisms involved the expulsion of a protonated N(b) followed by readdition.

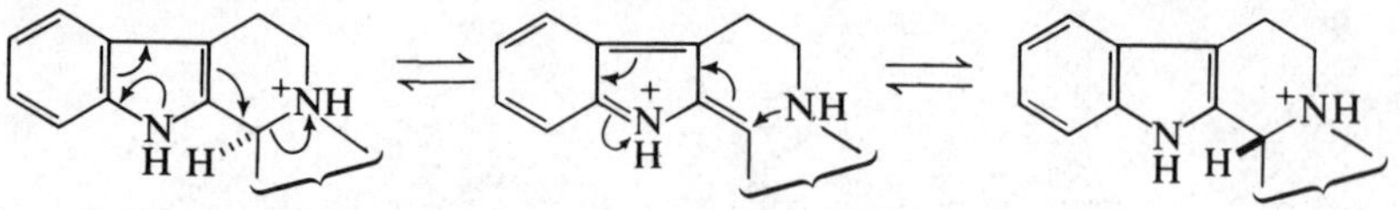

But, as the molecule (**13**) stands, inversion at C-3 was energetically unprofitable. The best conformation (XII) for **13** has all the groups in ring E equatorial, whereas with the desired C-3 epimer of **13** (XIII) they are all axial.

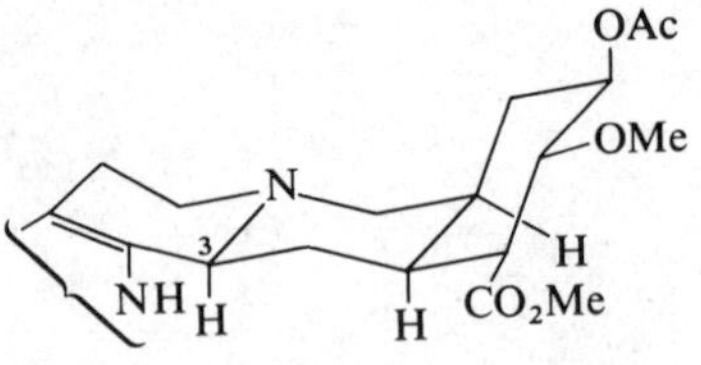

(XII)

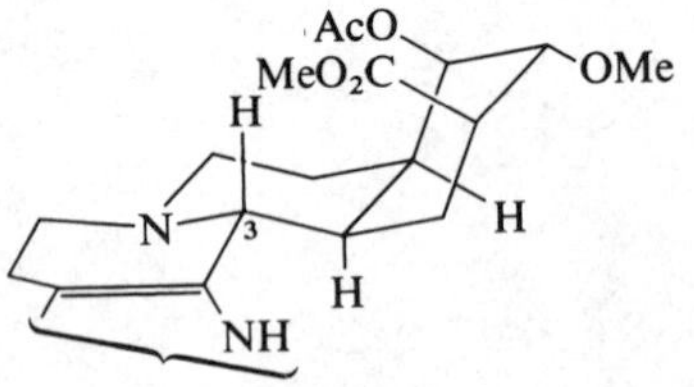

(XIII)

The ingenious solution to this problem was to hydrolyse both ester groups of **13** and make a lactone across ring E, using dicyclohexylcarbodiimide (cf. p. 84 and p. 195). The best conformations for **14** and its C-3 epimer are those shown in XIV and XV.

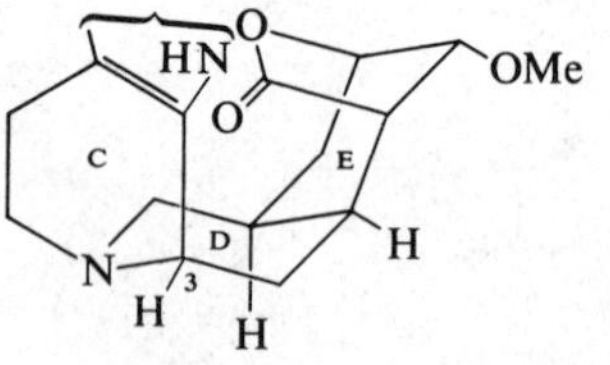

(XIV)

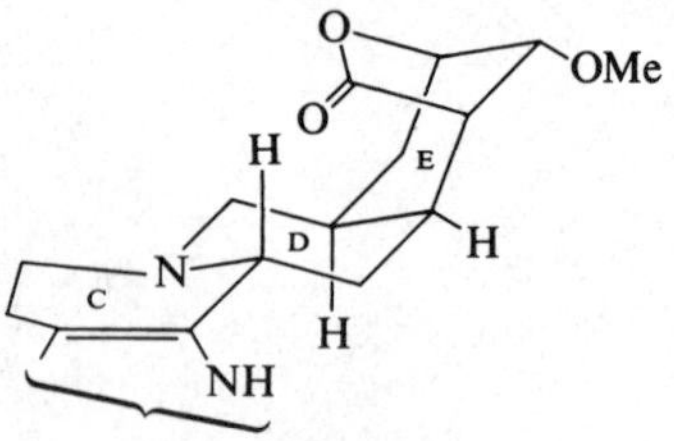

(XV)

Since the substituents in ring E were now necessarily axial, the normally flexible *cis*-decalin type of system in rings D and E was locked; the result was that, if chair conformations were to be maintained, the indole ring joined to ring C must be axial in XIV (i.e. **14**) but would be equatorial in XV. Thus the order of stability was reversed; and it was indeed found that, on heating with pivalic acid, **14** isomerised at C-3 to give **15** (i.e. XV). Methoxide in methanol opened the lactone, and acylation with 3,4,5-trimethoxybenzoyl chloride gave racemic reserpine, which was resolved using camphor-10-sulphonic acid.

Questions

1. What other reasonable mechanisms can you suggest for the inversion at C-3 of tetrahydro-β-carbolines using acid? Why do you think pivalic acid was used and not, say, acetic acid, which would have been more customary?

2. The periodic acid oxidation releases an aldehyde group which would, until the first borohydride reduction, be very likely to be capable of causing epimerisation at the adjacent chiral centre. Why do you think that this was no cause for worry?

References

1. R. B. Woodward and W. E. Doering, *J. Amer. Chem. Soc.*, **67**, 860 (1945).
2. R. B. Woodward, M. P. Cava, W. D. Ollis, A. Hunger, H. U. Daeniker and K. Schenker, *Tetrahedron*, **19**, 247 (1963).
3. R. B. Woodward, F. E. Bader, H. Bickel, A. J. Frey and R. W. Kierstead, *Tetrahedron*, **2**, 1 (1958).
4. E. C. Kornfeld, E. J. Fornefeld, G. B. Kline, M. J. Mann, D. E. Morrison, R. G. Jones and R. B. Woodward, *J. Amer. Chem. Soc.*, **78**, 3087 (1956).
5. R. B. Woodward, *Pure Appl. Chem.*, **6**, 561 (1963).
6. R. B. Woodward, K. Heusler, J. Gosteli, P. Naegeli, W. Oppolzer, R. Ramage, S. Ranganathan and H. Vorbruggen, *J. Amer. Chem. Soc.*, **88**, 852 (1966); R. B. Woodward, *Science*, **153**, 487 (1966); see p. 85.
7. R. B. Woodward, *Pure Appl. Chem.*, **2**, 383 (1961); see p. 112.

Index

Because the main aim of this book is to illustrate, in context, the reactions of organic chemistry, the index is primarily an index of reagents, reactions, and reaction types. Page numbers in *italics* indicate that the item referred to will be found on that page—usually in the structural diagrams—but that the word itself will not be found in the text: thus an example of an esterification reaction which is not specifically described as esterification will be found in the index, but the page number will be in *italics* and the esterification reaction will usually be found in the structural diagrams. Likewise, examples of stereoselectivity not remarked upon in the text as such are entered in the index with an italic page number. This system is intended to simplify the task of searching for examples of any particular kind of reaction or the use of any particular reagent.